Luciane Scheuer

Seasonality of Tourism in the Municipality of Guaratuba-PR, Brazil

Luciane Scheuer

Seasonality of Tourism in the Municipality of Guaratuba-PR, Brazil

Seasonality and its impacts

ScienciaScripts

Imprint

Cover image: www.ingimage.com

This book is a translation from the original published under ISBN 978-613-9-62133-0.

Publisher:
Sciencia Scripts
is a trademark of
Dodo Books Indian Ocean Ltd. and OmniScriptum S.R.L publishing group

120 High Road, East Finchley, London, N2 9ED, United Kingdom
Str. Armeneasca 28/1, office 1, Chisinau MD-2012, Republic of Moldova, Europe
Printed at: see last page
ISBN: 978-620-7-71834-4

SUMMARY

I dedicate this work to the residents of Guaratuba, who suffer annually from the impacts of seasonal tourism.

ACKNOWLEDGMENTS

To my mother, Sirlei, a wonderful person who was by my side at all times, trusting in me and my potential.

To my daughter, Leticia, who understood the importance of my absence and was an encouragement on the most difficult days.

To my father, Rudi, a man who always considered education to be the most important inheritance a father can leave to a son and believed in my dreams.

To my brother Alex, who always listened to me and gave me advice in the most difficult times.

To Our Lady of Aparecida, who has certainly illuminated my paths.

To my beloved husband, Igor Schott, for his support in the most difficult times and for understanding my absences.

To my father-in-law, Daniel Schott and my mother-in-law, Terezinha, for being my family in Guaratuba.

Beatriz and Jair Pedralli, great friends, who welcomed me into their home throughout my master's degree, without ever asking for anything in return.

To those who encouraged me to study Geography and to start in this field that is so essential to the study of the human sciences: Leandro Fontoura (Mineiro), Marciel Lohmann and Camila Cunico.

To my advisor, Prof. Dr. Miguel Bahl, whose patience, wisdom and understanding made a great dream possible.

To Prof. Dr. Marcos Aurélio T. da Silveira, who always encouraged me and believed in my potential.

To Prof. Dr. Alexandre Biz, who made his time available to participate in the qualification and defense of this study.

To my beloved colleagues: Valéria Albach, Eduardo Hack Neto, Leonardo Ravaglia Gonçalves, Simone Ramos and Vinicius Bonelli for the partnership, the good conversations, the friendship in class, at events, in articles and in life.

To Guaratuba City Hall.

Without effort, persistence and generosity, human beings will hardly get anywhere. (Author Unknown)

SUMMARY

The growing importance that tourism has assumed in global terms and particularly within the framework of some Western economies that are becoming less and less competitive in terms of traditional economic activities, justifies the great attention of the public and private sectors, both in terms of obtaining a more in-depth knowledge of the tourist phenomenon and, more specifically, in terms of the constraints inherent in its affirmation and progress. And within this phenomenon comes seasonality, which is something very complex and its study can identify its impacts, its problems and rationalize solutions that are appropriate in each situation or receiving community. Seasonality can be based on various causes, but this study makes it clear that climatic and geographical factors should not be seen as the sole culprits of seasonality. The problem can be triggered by a number of factors arising from a variety of reasons, both at the receiving and emitting centers. This study sought to characterize and prove the existence of seasonal tourism in the municipality of Guaratuba - PR, by approaching concepts related to tourism and the geography of tourism and turning them towards the study of seasonal tourism. It also sought to characterize in detail the object of study, namely Guaratuba. Finally, the study questioned 11 people directly and indirectly involved in tourism in the municipality, through focused interviews for later analysis and proof of the phenomenon. From the combination of the theoretical review and the analysis of the data from this exploratory study, it was possible to characterize and prove the existence of seasonality in the municipality of Guaratuba - PR, its causes and its effects.

Keywords: Tourism. Guaratuba. Causes. Effects. Seasonality.

INTRODUCTION

Studying Guaratuba is, antagonistically, a great challenge and a pleasure. The challenge lies in carrying out research into tourism in a city that suffers annually from the impacts of seasonality and a lack of tourism planning, located in one of the regions with the greatest diversity of landscapes in Paranà and with a unique historical, economic and cultural context. It is a city that daily experiences the desires of those who want it to be big and beautiful, and the difficulties of those who build and manage it, in a constant conflict that spatializes itself over time, forming territories and territorialities that are difficult to understand. For this reason, it is a challenge to study a city that has a considerable lack of data, either because of the scarcity of studies and research or because of the lack of incentive on the part of public bodies to organize it properly. And it's a pleasure to be able to analyze tourism in the city where I chose to live, work and write my story.

Studying Guaratuba is also rewarding because it means fulfilling a long-held desire to understand the dynamics that drive it, with its advances and setbacks, and in some way to be part of its current local reality.

This research looked at concepts related to tourism and the geography of tourism and turned them to the study of seasonal tourism, in order to characterize and prove the existence of seasonal tourism in the municipality of Guaratuba, on the coast of the state of Paraná. For Butler[9] (1994, p. 332 *apud* ALMEIDA; KASTENHOLZ, 2009) seasonal tourism is:

> A temporal imbalance in the tourism phenomenon, which can be expressed in terms of dimensions such as: number of visitors, visitor spending, traffic on highways and other forms of transportation, employment and entrance fees to attractions (ALMEIDA; KASTENHOLZ, 2009, p. 4).

Seasonality can therefore be considered to be the concentration of tourist flows in short periods of the year, promoting, on the one hand, peaks in activities providing general services and tourists, which often constitute a heavy burden on physical and social resources in the destination area and, on the other hand, seasons of low demand that generate inefficiency in local and regional tourism activity.

Summer resorts, such as Guaratuba, need proper tourism planning in order to reverse the situation of seasonality, so that tourism can be satisfactorily exploited all year round. According to Almeida (2000), the damage caused by seasonal tourism in many towns is much greater than the return it brings in season. "All it takes is one bad season, with bad weather, or problems in the national economy, and it will take years for the place to recover." (ALMEIDA, 2000, p. 73).

---

9 BUTLER, R. **Seasonality in tourism:** Issues and problems. Tourism: The State of Art. A. V. Seaton. Chichester: Wiley, p. 332-339, 1994.

The main idea of this study was to really build a theoretical study on concepts related to tourism and seasonality, an in-depth characterization of the object of research, i.e. Guaratuba and its peculiarities, and a field study, where it was possible to characterize and prove the existence of seasonality in Guaratuba.

The development of tourism in the municipality of Guaratuba, on the coast of the state of Paranà, is well-established, attracting a large number of tourists, especially during the summer vacation months. But what about the other months of the year, how do tourism and other economic activities behave in the municipality? Does the seasonality of tourism in Guaratuba have an impact on the quality of tourist services and on the life of the community in general? Based on the above problem and the need to study seasonal tourism in the municipality of Guaratuba, the following hypotheses were formulated: does seasonality occur due to a lack of tourism planning or due to climatic and territorial factors? Does seasonality occur due to a lack of commitment on the part of the business community to create reasons (events, programs, new attractions) or due to a lack of infrastructure? Or does seasonality lead to a drop in the population's income, causing economic and social problems?

As for the nature of the research, this study is original, because as Ruaro specifies:

> Research is considered original when it is a study carried out for the first time, i.e. the focus, the delimitation of the topic has not yet been approached from the aspect being presented. Thus, original research should contribute new discoveries to the expansion of scientific knowledge (RUARO, 2004, p. 24).

In other words, the research carried out in Guaratuba, in its approach to seasonality, took place for the first time. In terms of objectives, this research can be considered exploratory and descriptive, because according to Gil:

> The main purpose of exploratory research is to develop, clarify and modify concepts and ideas, with a view to formulating more precise problems and researchable hypotheses for subsequent studies. Of all the types of research, they are the least rigidly planned. They usually involve bibliographical and documentary surveys, non-standardized interviews and case studies. Sampling procedures and quantitative data collection techniques are not usually applied in these surveys (GIL, 2008, p. 27).

The main objective of this study was to analyze seasonal tourism in the municipality of Guaratuba - PR and its effects on the local community, in order to characterize and prove the existence of seasonality. The specific objectives are: to research bibliographic and documentary data on the proposed topic; to analyze the municipality in question geographically, socially, touristically and economically in order to characterize it; to identify and diagnose the municipality's strengths and weaknesses in relation to its seasonal and annual tourist supply and demand; to analyze the municipality's tourist demand; to survey important territorial elements for the study of tourism and local geography; to characterize and prove the existence of seasonality.

According to Dencker (1998), this type of research is suitable for the phases of reviewing the literature, formulating the problem, raising hypotheses, identifying and operationalizing the variables.

According to Gil, this research is descriptive because:

> The main objective of research of this type is to describe the characteristics of a given population or phenomenon or to establish relationships between variables. There are countless studies that can be classified under this heading and one of their most significant characteristics is the use of standardized data collection techniques (GIL, 2008, p. 28).

Dencker (1998) argues that descriptive research is indicated to guide the form of data collection when the aim is to describe certain phenomena.

The research was carried out using bibliographic and documentary research and intensive direct observation, which according to Marconi and Lakatos (2008, p. 75) "is carried out using two techniques: observation and interview".

The bibliographic and documentary research was carried out by means of indirect documentation, because according to Marconi and Lakatos "it is the phase of research carried out with the aim of gathering prior information and is done in two ways: documentary research (or primary sources) and bibliographic research (or secondary sources)." (2008, p. 48). This study used documents and publications from bodies such as the Instituto Paranaense de Desenvolvimento - IPARDES, the Instituto Brasileiro de Geografia e Estatistica, the Ministério do Turismo and the Prefeitura Municipal de Guaratuba, among others, characterizing the research as documentary and bibliographic, because according to Marconi and Lakatos "it covers all bibliography already made public in relation to the subject of study, from single publications, bulletins, newspapers, magazines, research [...]"...]". (2008, p. 57).

Based on the bibliography analyzed, the appropriate comparisons were made with what was seen in the field, that is, using field research, because according to Marconi and Lakatos (2008, p. 69), "it is that used with the objective of obtaining information and/or knowledge of a problem, for which an answer is sought [...]". The same authors also mention that "it consists of observing facts and phenomena as they occur spontaneously, collecting data about them and recording variables that are presumed to be relevant, in order to analyze them." Therefore, it can be said that this research was also characterized as direct documentation, which "generally consists of collecting data in the very place where the phenomena occur." (MARCONI; LAKATOS, 2008, p. 69).

Within the empirical part, this research is characterized as Intensive Direct Observation, through the two techniques used: observation and interview. The type of observation used in this research was asystematic observation, i.e. according to Marconi and Lakatos (2008) also called unstructured, spontaneous, informal, simple, free, occasional, which consists of collecting and recording the facts of a reality without the researcher using special technical means or having to ask direct questions. It is most commonly used in exploratory studies and does not have previously elaborated planning and control. The interview used was a focused interview, which according to Gil (2008, p.112) "is just as free as the informal interview, but focuses on a very specific topic. The interviewer allows the interviewee to speak freely on the subject, but when the interviewee deviates from the original topic, they make an effort to return to it." The type of sampling chosen was intentional non-probability, which according to Marconi and Lakatos (2008), is the most common type of non-probability sample, and in which the researcher is interested in the opinion of certain elements of the population, but not representative of it.

In the case of this research, the sample would be made up of 30 people directly or indirectly involved in local tourism. Of the 30 people contacted, only 11 were willing to answer the interview.

In this research, the author observed the behavior of tourists in high and low season and also the environmental impacts caused by mass tourism during the summer season, holidays and weekends.

As for the method of approach, this research can be characterized as hypothetical-deductive, because according to Gil:

> When the knowledge available on a given subject is insufficient to explain a phenomenon, a problem arises. To try to explain the difficulty expressed in the problem, conjectures or hypotheses are formulated. From the hypotheses formulated, consequences are deduced that must be tested or falsified (GIL, 2008, p. 12).

With regard to the method of procedure, this study is characterized as a monographic or case study method, as it "starts from the principle that the study of one case in depth can be considered representative of many others or even of all similar cases. These cases can be individuals, institutions, groups, communities, etc." (GIL, 2008, p. 18).

Finally, the answers obtained were interpreted and analyzed so that the seasonality of tourism could be characterized and proven in the municipality of Guaratuba.

The results of this study are of great relevance for the development of other studies on tourism in general, for geography, as it takes into account various geographical aspects of a territorial, spatial and economic nature, among others, for people interested in this subject and especially for the municipality of Guaratuba, which suffers the effects of seasonality every year. In this way, we can see that places that live with seasonality must try to develop strategies to deal with it, in order to minimize the situations of high and low flows.

This study is divided into four chapters: Chapter I reviews the literature on tourism, territory and tourist space, first conceptualizing tourism, analyzing the evolution of tourism in the world and focusing on the importance of tourism geography for the study. Secondly, an attempt was made to analyze the relevance of the tourism system and its main subsystems and their territorial articulation. Subsequently, some theories on tourist space were discussed, from the point of view of various authors, as well as the structuring of space in coastal areas. From there, arguments were made about mass tourism, second home tourism and tourism planning.

The second chapter deals with seasonality, definitions, causes and effects, seasonality in coastal areas and its impacts.

Chapter III sought to characterize the object of study, i.e. the municipality of Guaratuba, starting with some arguments about the coast of Paraná, the characterization of the municipality in various aspects, with various analyses of the data collected.

Lastly, Chapter IV deals with the work methodology and the field research, carried out by means of focused interviews and systematic observation, for subsequent data analysis and final considerations.

CHAPTER I - TOURISM, TERRITORY AND TOURIST SPACE

The conceptualization of tourism and the contents that follow have been drawn up with the intention of demonstrating the diverse and complex nature of the tourist phenomenon and its geographical aspect. Tourism, which is based on the social valorization of certain geophysical and cultural components, appears as a type of recreational activity, or not, which takes place on a geographical scale, fundamentally local, regional, national or international. The difference with other recreational activities of a domestic nature is that tourism is characterized by leaving the place of habitual residence for a day or more.

1.1 CONCEPTS OF TOURISM

Human beings have always moved around for a variety of reasons, such as the search for food or shelter, the domination of territories, the conquest of wealth, as well as religious, cultural and social factors, among others. However[1] "the displacement of people known as the phenomenon of tourism, which is a social practice with economic characteristics, only began to develop in the 19th century, more precisely after 1860." (BONFIM, 2007, p. 39).

Great mobility has taken hold of most of the inhabitants of industrial nations. They take every opportunity to travel. Opportunities to get away from everyday life as often as possible. Short breaks during the week or weekend, long trips on vacation. For many people there is no greater desire than to travel, to change address, which is why the study of tourism is so important and complex.

Over the decades, tourism has experienced continuous growth and deepening diversification to become one of the fastest growing economic sectors in the world. Modern tourism is closely linked to development and encompasses an increasing number of new destinations. This dynamic has turned tourism into a kind of "essential engine" for the socio-economic progress of various localities.

According to the World Tourism Organization - WTO (1995), a tourist is a person who travels to a place other than their home for a period of less than 12 months and whose main objective is not to carry out a paid activity in the place visited. When people are tourists, they carry out activities during their trips and stays in different places from their usual homes, for a period of less than a year, for leisure, business or other reasons not associated with paid activities in the place visited.

This concept reflects a world reality that is not yet globalized. Accompanying the global changes underway, the UNWTO (2001, p. 38) proposes a new definition in the following terms: "Tourism comprises the activities carried out by people during their travels and stays in places other than their usual environment, for a consecutive period of less than one year, for leisure, business or other purposes."

As it is a complex and widely expanding subject, there are as many definitions as there are authors writing about it. For Barretto:

> Tourism is essentially the movement of people and meeting their needs, as well as the needs of other people who don't travel. Tourism is the phenomenon of interaction between the tourist and the receiving center and all the activities resulting from this interaction (BARRETTO, 1991, p. 47-48).

It is also worth noting another conceptualization from Silveira (2002), who defines tourism as:

> [...] an activity that consists of the displacement of people, temporary and limited in time and space, in such a way that it does not imply the transfer of the habitual place of residence, which has diverse motivations (which can range from simple leisure, through curiosity, education, health, culture, adventure, to professional and religious aspects), and which has, on the one hand, a desire to escape from one's daily surroundings, and, on the other hand, the search for new spaces and cultures in a more or less linked way, which has economic, social, cultural and environmental effects. Tourism is therefore seen as an activity that also produces and consumes spaces, and is responsible for new territorialities [...] (SILVEIRA, 2002, p. 21).

Tourism can therefore be considered a multidisciplinary activity which, in its breadth, is not only made up of the people who travel and the goods and services offered and used, but also of a series of relationships and situations of an economic, sociological, geographical and even political nature.

Bahl (2004, p. 45), adds that,

> Essentially, tourism can be a promoter of the localities where it takes place, forcing the need to rescue local culture, stimulate self-esteem and value local habits and customs. Maintaining the authenticity of localities is one of tourism's most promising resources (BAHL, 2004, p. 45).

Some scholars treat tourism as an industry: Fullana and Ayuso[10] , cited by Dias (2003, p. 10),

> An important characteristic of the tourism industry is that, unlike manufacturing or other service industries, it doesn't manage most of the products and experiences it sells. Instead, tourism managers transport individuals to get to know the natural features, cultural attractions and lifestyles of a tourist destination. (DIAS, 2003, p.10)

For his part, Beni (1998, p. 36) identifies three trends in the definition of tourism in the academic field, in companies and in government bodies: "the economic, the technical and the holistic." The definition written by Robert McIntosh[11] (1977, *apud* BENI 1998), states that:

> Tourism involves more than business components, presenting a qualitative facet, it can be defined as the science, art and activity of attracting and

10 FULLANA, P. and AYUSO, S. **Turismo Sostenible**. Barcelona: Rubes, 2002
11 MCINTOSH, E. R. **Tourism principles, practices, philosophies**. Ohio:Columbus, 1977.

transporting visitors, accommodating them and courteously satisfying their needs and desires (BENI, 1998, p.36).

As for the technical concepts related to tourism, these are practically linked to the definition of tourists, which, according to Beni (1998, p. 37), based on the idea of the United Nations (1963), states that, "they are temporary visitors who stay at least twenty-four hours in the country visited, whose purpose of travel can be classified under one of the following topics: leisure, business, family, missions and conferences."

And finally, the holistic definition seeks to encompass the total essence of the subject.

Jafar Jafari (*apud* BENI, 1998, p. 38) argues that: "It is the study of man far from his place of residence, of the industry that satisfies his needs, and of the impacts that both he and the industry generate on the physical, economic and socio-cultural environments of the receiving area."

But as tourism is a really broad activity, it is worth mentioning other characteristics that are pertinent to its study. Tourism can be said to be an efficient means of:

1. to promote the dissemination of information about a particular region or locality, its natural, cultural and social values;
2. opening up new social perspectives as a result of the region's economic and cultural development;
3. integrate socially, increase (in certain cases) national awareness;
4. developing creativity in various fields;
5. promoting a sense of freedom by opening up to the world, establishing or extending cultural contacts and stimulating interest in tourist travel (BENI, 1998, p. 41).

But as an economic activity, tourism can also cause damage and harm to the environment it visits. For Beni, the damage is as follows:

1. degradation and destruction of natural resources;
2. loss of the authenticity of local culture;
3. stereotyped and false description of the tourist and the country or region they come from, due to lack of adequate information;
4. lack of prospects for those groups of the local population in tourist destination areas who do not benefit directly from tourist visits or from the local tourism system itself;
5. the appearance of phenomena of social dysfunction in the family, pathology in the process of socialization, disintegration of the community;
6. dependence on foreign capital or existing stereotypes about tourism (BENI, 1998, p. 41).

According to the UNWTO (2010), the turnover of tourism is currently equal to or even exceeds that of oil exports, food products or automobiles. Tourism has become one of the main players in international trade and at the same time represents one of the main sources of income for many developing countries. This growth is accompanied by greater diversification and competition between destinations.

This global spread of tourism in industrialized and developed countries has produced economic benefits in various sectors related to construction, agriculture and telecommunications.

For all these reasons, the conceptualization of tourism cannot be limited to a simple definition, as it is a phenomenon that occurs in different fields of study, is explained according to different currents of thought, and is verified in various contexts of social reality.

In view of the concepts cited as a basis for gaining a better understanding of the real meaning of tourism, it can be seen that tourism is a broad and indeed complex phenomenon. Thus, based on the various approaches presented, it is pertinent to mention that, for this dissertation, the concept of Barretto (1991, p. 47-48) was adopted as a basis, which states that "tourism is a phenomenon of interaction between the tourist and the receiving center and all the activities resulting from this interaction." This concept is in line with the discussions that will follow in the course of this work.

1.2 THE EVOLUTION OF TOURISM IN THE WORLD

Tourism began to be organized in the mid-19th century, but it was not until the 1950s that it became a significant mass activity in socio-economic and cultural terms (KRIPPENDORF, 1989).

Trigo (1993) argues that the major tourism movements in the 19th century emerged as capitalism developed to an industrial level, especially in Western European countries and North America. Various consequences of this second phase of capitalism (first phase: commercial or mercantile capitalism) marked the 19th century and helped to strengthen tourism. According to Trigo (1993, p. 15), the main ones were "technological research, which made it possible to build steam engines to equip ships, locomotives and industrial production machines, and new engineering and architectural resources that revolutionized civil construction".

Tourism was born and developed with capitalism. Because it is a tertiary sector activity (providing services) and because it is not considered a basic necessity, it has suffered, and still suffers, from all the crises in history. But it was from the 1960s onwards that tourism emerged as a source of pleasure for millions of people and as a source of profit and investment, with guaranteed *status* in the world of international finance.

For wheat:

> Contemporary tourism influences and is influenced by a broader context that encompasses the economy, society, politics and culture. As far as relations with culture are concerned, it should be borne in mind that the profound changes that have taken place in recent decades in education, the arts and general culture itself, require a well-defined perception of what this post-modern culture that permeates post-industrial societies is (TRIGO, 1993, p. 36).

Contemporary culture reserves a privileged space for pleasure and leisure. Therefore, understanding cultural issues is important in order to understand the implications for these fields, especially tourism, which is the special objective of this study.

And in this evolving context of tourism, we can see that its study is part of the study of the social practices carried out by human beings in the process of organizing space, a process that also falls within the remit of geography. According to Corrêa (2000), throughout the long process of organizing space, human beings have created, maintained, undone and redone different forms of spatial interaction, in order to alter space, partially or totally. It is the human being's awareness of spatial difference that gives rise to social practice.

Corrêa argues that:

> [...] social practices result, on the one hand, from man's awareness of spatial differences. This awareness is anchored in cultural patterns specific to each type of society and in the technical possibilities available at any given time, which give different meanings to previously differentiated nature and spatial organization [...]. On the other hand, they result from the various projects, also derived from each type of society, which are devised to enable the existence and reproduction of an activity or a company, a specific culture, ethnic or religious, for example, or society itself as a whole (CORRÊA, 2000, p. 35).

According to Krippendorf (2000), industrial nations are mainly responsible for the development of tourism as an economic activity, as their inhabitants need to change their places of residence and production for others that offer them freedom, leisure, happiness, rest and security. Tourism was born and developed with capitalism, establishing itself as a source of profits and investments, with significant performance in the international financial world, generating jobs and playing a significant role in the foreign exchange earnings of different regions of the world.

1.3 THE GEOGRAPHY OF TOURISM

Tourism is, in essence, a social practice. A social practice, added to the market, which has space as its main object of consumption and, as a result of this intrinsic characteristic, requires the adaptation of territories to its material and immaterial demands. According to Crocia:

> Tourism usually expands into areas where there used to be human settlements. This means that when the diffusion of the tourist function begins to take place in

a region, an interaction begins between the components of the new function (tourism) and the components of the receiving region. (CROCIA, 2002, p. 9).

It can even be said that the previous qualities presented by a region end up influencing, and often decisively, the very decisions of private and public agents to invest in tourism in that area. Pearce[12] (1988, *apud* CROCIA, 2002, p. 13) observed that, in general, tourism develops within a previous economic structure and, as such, "certain forms of urban hierarchy and transportation networks already exist". In addition, Pearce[13] (2003 *apud* CROCIA, 2002, p. 13) returned to the subject and emphasized the importance of using a more regional and integrated approach when analyzing tourism from a geographical perspective, in order to make the relationships between the nodes of urban locality networks evident.

The work of Mitchell and Murphy[14] (1991 *apud* BONFIM, 2007, p. 44) defines the object of study of Tourism Geography as the interrelationships that are created between tourists, the local population and the activities carried out, with space as the reference. In this context, studies relevant to the Geography of Tourism would address environmental, regional and evolutionary aspects of space. According to Boniface and Cooper[15] (1994 *apud* BONFIM, 2007, p. 44), the object of study of Tourism Geography is the spatial expression of Tourism as a human activity, centered on both the area of origin and destination, as well as the relationship between the two.

Pearce[16] (1988 *apud* VERA, 1997, p. 31) indicates six areas that make up the Geography of Tourism: a) the patterns of spatial distribution of supply; b) the patterns of spatial distribution of demand; c) the geography of vacation centers; d) tourist movements and flows; e) the impact of tourism; f) models for the development of tourist space.

In other words, it can be stipulated that the study of tourism development covers all these areas, prioritizing issues relating to supply, location and impact. Ivars (2002, p. 19) mentions the existence of two lines of work that cover the object of study of Tourism Geography, "the perspective of spatial analysis in its applied dimension, which is very evolved with respect to traditional formulations, and the more reflective and critical approaches, which expand and update the weak theoretical body of Tourism Geography."

Although the lines are different, they are not contradictory and are undoubtedly of interest to tourism planning. In Brazil, according to Rejowski (1996), the study of tourism has been

---

12 PEARCE, D. **Desarrollo Turistico:** su planificăcion y ubicación geografica. Mexico: Trillas, 1988.
13. **Geography of Tourism:** flows and regions in the travel market. Sâo Paulo: Aleph, 2003.
14 MITCHEL, L.; MURPHY, P. Geography and tourism. **Annals of Tourism Research**, n. 18, p. 57-70, 1991.
15 BONIFACE, B. G.; COOPER, C. **The geography of travel & tourism**. Oxford: Butterworth-Heinemann, 1994.
16 PEARCE, D. **Tourism development**. Its planning and geographical location. Mexico: Trillas, 1988. p.
14.

evolving with efforts in research and teaching in a similar way to the scientific process that has already taken place in other areas of the human and social sciences, outlining parameters with their characteristic components and dimensions for tourism. Bonfim (2007, p. 45) mentions that research has shown that "it has been very difficult to adapt Tourism and its contents to the needs of the Geography course, as well as the absence of a theoretical-methodological framework that indicates the best way to do this." This is because the paradigms of Tourism Geography have changed over time, and the evolution of the field of analysis is taking a course that shows the existence of different approaches.

Vera (1997) points out that:

> [...] in the same period we find different ways of doing tourism geography in different countries, for example, in the UK, France, Germany and the USA, we refer to 'cultural' areas, which are in turn differentiated, with the predominance of some themes over others and the use of contrasting approaches and methodologies. Thus, in a very simplified way, we can say that in Germany priority is given to morphological (the landscape) and social aspects; in France, to international tourism and models for small and large scale, and in the United States and the United Kingdom, although with differences, to recreation in rural and natural areas. (VERA, 1997, p. 29).

According to Vera (1997), among the scientific productions are those that rely on a tourism system as a methodological basis for tourism geography. The operational aspects of Tourism Geography are the planning and management of tourist territories.

After indicating these references, it is necessary to emphasize that the different methodological lines should not be considered according to the Cartesian philosophy, as the contribution of each can certainly reveal a richer and more complex reality. Bonfim (2007, p. 52) starts from the premise that:

a) Geography, as a science, researches, analyzes, interprets and synthesizes the way in which human beings, throughout their existence, have occupied, used and transformed nature and also organized space according to their needs and desires;

b) Tourism involves the consumption, production and transformation of natural objects into social ones.

In conclusion, the reality of the tourist area can only be understood through knowledge and understanding of certain geographical concepts. Consequently, future tourism professionals must master these concepts.

According to Coriolano and Silva (2005, p. 23), "one of the concepts that must be discussed, assimilated and applied to tourism is the concept of geographical space: nature modified by man through work".

We should conclude by saying that there is currently a diversity and pluralism in the analysis of activities related to leisure and tourism, which has been accentuated since the 1970s, when eclecticism in the social sciences became more relevant. The Geography of Tourism has emerged as a way of studying the territories, places, spaces, among other aspects, where the tourist phenomenon takes place and a new social dynamic and undoubtedly new tourist behaviors.

1.4 THE TOURISM SYSTEM

The social sciences have used systems theory to recognize complex processes in which societies, territories and economies are articulated and which together form a functional whole. According to Beni:

> A system can be defined as a set of parts that interact in order to achieve a certain end, according to a plan or principle; or a set of procedures, doctrines, ideas or principles, logically ordered and cohesive with the intention of describing, explaining or directing the functioning as a whole (BENI, 1998, p. 25).

In other words, examining and reflecting on reality from a methodological perspective, which understands it as a cog that works by adding up its parts. According to Thornes and Brunsden[17] (1977 *apud* BENI, 1998, p. 25) the system is seen as "a set of objects or attributes and their relationships, which are organized to perform a particular function". From this perspective, the system is an operator which, over a given period of time, receives *input* and transforms it into *output*. Remembering that *input* means input and *output* means result. In other words, the system is a complex functional chain that can be atrophied or fractured if any of its parts do not behave systematically, which can happen due to a lack of *input,* its own pathologies or a lack of functional behavior with the whole. Vera (1997) argues that:

> The systemic approach applied to tourism is necessary because of the diversity of the components that shape, produce and determine the dynamics of tourism as a whole and demands an effort of scientific rationalization and methodological approach, always conscious of simplifying the reality of things (VERA, 1997, p. 38).

Within the tourism processes, Vera (1997, p. 39) cites at least three vertices of a possible systemic articulation: the tourist as the final object of the whole system, the agents who design and organize the tourism products that compete in the markets and tourist spaces, and the mobility of the trip as the articulation between sending and receiving.

For Beni (1998), the Tourism System is also called Sistur and as the author argues, the general objective of Sistur is:

> To organize the study plan of tourism activity, taking into account the need, long demonstrated in theoretical works and research published in various countries,

---

17 THORNES, J.B., BRUNSDEN, D. **Geomorphology on time**. London: Methuen, 1977.

> to substantiate working hypotheses, justify scientific positions and principles, perfect and standardize concepts and definitions, and consolidate research approaches to provide tools for analysis and expand research, with the consequent discovery and development of new areas of knowledge in tourism. (BENI, 1998, p. 44).

In this way, we can see that Sistur is an open system, because it exchanges with its surroundings and is therefore interdependent, not characterized by static structures and functions. Precisely because it is open, it maintains a continuous process of dialectical relationships of conflict and collaboration with its surroundings. Beni goes on to say,

> In theory, therefore, the system would have to exploit its tourist attractions in such a way that a degradation - through use - would be matched by an opposite force, driven by man, of a conservative or regenerative nature, in order to guarantee the system's long-term durability, such as, for example, a city with a coastal tourist vocation. (BENI, 1998, p. 51).

For Vera (1997), the tourism system has four elements, but before highlighting what these elements are, it is important to identify some unique and constitutive aspects of the tourism dynamic.

> In any case, we must repeat what is already sufficiently well known: firstly, tourism consumption and production take place simultaneously, as the service industry that it is, and secondly, demand must appreciate it, and not the other way around, as is the case with the manufacturing industry, where there is the resource of tourist attraction (natural, cultural, monumental, sporting...), converted by travel and tourism operators into a tourist product (VERA, 1997, p. 40).

In this respect, the analogy of considering the tourist resource as the raw material of the process has often been used, while the tourist product would be the good already prepared and ready for consumption. So what are the elements that make up the tourism system? According to Vera, these elements are

> a) Tourists as consumers who decide, and are *already* conditioned in their decision by their socio-demographic profile, the media and tour operators...; b) The public and private mechanisms that promote and market tourism resources, converting them into products, encouraging, creating and doing everything possible to satisfy tourists' motivations.... c) The transport, connectivity and tourist mobility systems that organize travel between the sending and receiving areas... d) The tourist destination or territory, which must be a more or less recognizable setting with great ability to attract tourist flows [...] (VERA, 1997, p. 40).

The question is whether the tourist area really works as a true system, whose gears, components and parts behave as a system functionally to satisfy tourists, obtain dynamic competitive advantages and constant improvements in the social, environmental and economic profitability of tourist destinations. Therefore, this deductive system must be applied empirically to the reality in which tourism production and territorialization relations are embedded.

The tourist system as a methodological tool has a spatiality that can make it possible to recognize more clearly the different tourist systems that exist in today's world and to understand their functional problematics, in a way that complements other purely economic, environmental or sociological approaches.

Boullòn (1997) proposes an explanatory version of the tourism system, which means considering that tourism is not made up of many systems but of various facets of a system and that the study of each of these facets is what has given rise to different analytical models, one of which is called "supply-demand". The other models cited by Boullòn (1997) are the social anthropological and what is known as industrial tourism.

> The supply-demand model focuses its interest on the functioning of so-called commercial tourism. The social anthropologist is concerned with the manifestations of leisure and free time in different societies and analyzes their repercussions on individual and collective behavior, while the industrial model is interested in mass production, commercialization and profit (BOULLÓN, 1997, p. 31).

Boullòn (1997) presents a figure that represents the integral parts of the tourism system according to the supply-demand model, as shown below:

FIGURE 1 - HOW THE TOURISM SYSTEM WORKS SOURCE: Adapted from BOULLÓN (1997, p. 32).

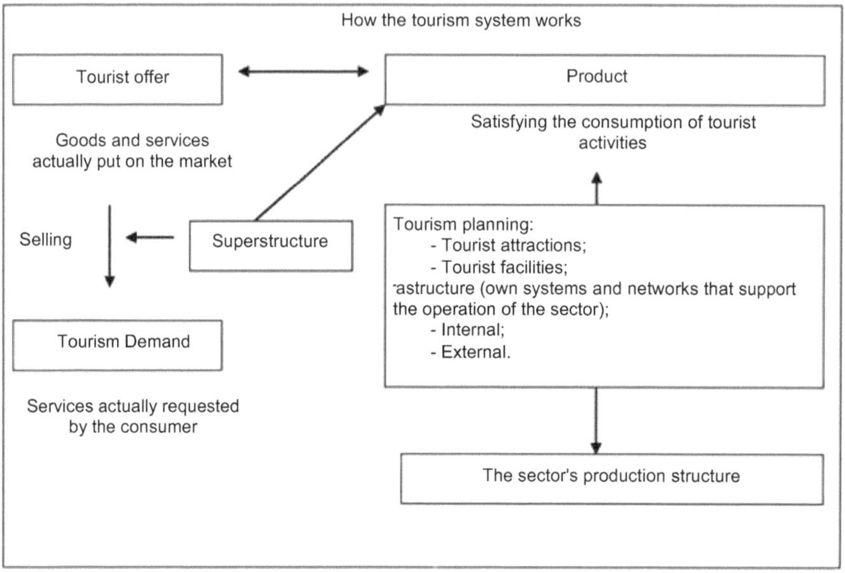

The figure can be explained as follows: on the left is the starting point of the system's operation, which originates in the meeting of tourism supply and demand through a process of selling the so-called tourism product which, together with the infrastructure, forms the production structure of the sector, as indicated on the right-hand side of the figure. In the center is the superstructure, whose function is to control the efficiency of the system, coordinating the functioning and interrelationship between the parts.

According to Boullón (1997), the components of the tourist system are: tourist demand, tourist supply, the sales process, the tourist product, the tourist site and attractions, the infrastructure, the superstructure and the tourist heritage. Unlike the other authors cited, Boullón (1997) tries to facilitate understanding of the tourism system in a more didactic and comprehensible way. The components of the tourism system according to Boullón (1997) are explained below, with his own arguments and those of other authors.

People who leave their usual homes temporarily, for recreational purposes or for other reasons or needs, demand the provision of certain services. Demand is therefore made up of individuals looking for goods and services in places other than their place of residence. For Beni:

> The satisfaction that tourists seek through the consumption of tourist goods and services involves the acquisition of very diverse experiences, found in the

consumption of different components of the product - the "utility function", according to conventional economic language (BENI 1998, p. 201).

Since travel is a movement between two points, tourism demand is the demand for this movement and depends on both the characteristics of these two points and the cost of this movement.

A market demand for a product, in this case tourism, can be stipulated as the "total volume that would be purchased by a defined group of consumers, in a given geographic area, in a defined period of time, in a defined market environment, under a given marketing program." (KOTLER, 1994, p. 220). It can be seen that total market demand is not just a fixed number, but a function of stated conditions.

Therefore, the demand for tourism is not made up of simple elements or specific services in isolation. Goods and services are demanded that complement each other. Man, the active subject of tourism, bases his preferences on the existence of multiple elements or conditions, which constitute the spatial factors of the tourist decision, which according to Beni (1998, p. 213) "are the nature of the resources, the environment, communication, climate, the action of man, relative size and personality." Mota (2001, p. 78) adds that "there are also temporal factors, which are fundamental for there to be demand and for the phenomenon of tourism to be realized." Among them is free time, so that the individual can allocate it to tourism activities.

According to Bote Gòmez (1990), the study of tourism demand is based on three groups of questions:

> a) socio-economic characteristics (gender, marital status, profession, level of education, income, residence, etc.); b) quantitative and qualitative characteristics specific to the trip (motivation, stay, type and characteristics of accommodation used, expenses, way of making the trip, etc.)); b) the specific quantitative and qualitative characteristics of the trip (motivation, stay, type and characteristics of accommodation used, expenditure, way of making the trip, transportation, attractions, among others); and c) visitors' opinions on certain aspects of the tourist activity (level of satisfaction and loyalty, opinion on prices and quality of services provided, etc.) (BOTE GÓMEZ, 1990, p. 116).

From the above, we can see the importance of studies on seasonality and demand characteristics for the strategic planning of tourism in a locality, since in order to meet the needs of the consumer, it is first necessary to know them in depth. It's important that this information, which is in the public domain, is really accessible for consultation by the population and, above all, by businesspeople who, by using it properly, can contribute to a significant increase in local, regional and national tourism demand.

For Wahab (1991), tourism demand can be classified into potential and actual. Potential demand can be understood as a part of the population that has at least leisure time, disposable income and a desire to travel, as well as other rational and irrational factors. Real demand, on the other hand, is represented by individuals who have actually traveled to a certain place. Mota (2001) argues that the quantitative difference between potential and actual demand is a field of dispute for marketing activities.

It is worth pointing out that the main characteristic of tourist demand is its heterogeneity (BENI, 1998, p. 202), "because its structural stratification does not follow a systematic model, based, for example, on socioeconomic, geographical, behavioral or other variables." An in-depth analysis of motivations is needed in order to better understand the phenomenon of tourism and offer products and services that meet the expectations of demand.

It can therefore be concluded that the study of the various factors that influence tourist demand and consumption is of great importance, since a person's purchasing decision is the result of a complex interaction of cultural, social, personal and psychological factors, some of which will not be influenced by marketing actions. According to Boullon (1997), without these studies it is very difficult to know what should be done or built to increase demand for products and services in tourist locations.

In an economic analysis, supply can be said to be the quantity of products or services that enter the consumer market at a given price and for a given period of time. In tourism, supply, according to Mota (2001, p. 69) "constitutes the raw material of the activity through a set of natural and cultural resources, to which tourist services such as transportation, accommodation, food and others are added, together forming the tourist product." According to Pierre Defert[18] (1996 *apud* BENI, 1998, p. 155-156):

> Primary tourist resources or the original offer is made up of water elements in all their forms (hydromo); natural surfaces covered with vegetation by the voluntary action of man and flora (ritomo); values created by the activity of man with history, religion, culture, folklore and others (litomo); and the ancient and modern activities of man (anthropomo). (BENI, 1998, p. 155-156).

This notion that natural and cultural resources are raw materials for tourism is important if we are to become aware of their content in the national tourism product and the importance of maintaining and conserving them through the correct use of these resources over time, in a way that fosters self-sustaining tourism. Mota (2001) points out that these elements are factors that differentiate the tourist product, making it a unique phenomenon, since it is impossible to faithfully reproduce natural and cultural environments that are different from their place of origin.

A second group of supply is made up of derived tourism supply, which according to Mota

---

18 DEFERT, P. **Economic structure and localization in tourist regions.** Paris, Institut International de Recherches Touristiques, 1996.

(2001, p. 70), "regroups all the services provided by tourism companies. This type of offer cannot satisfy demand unless there is a combination of various factors from the two types of offer - derived and original."

In order to assess the tourism potential of a region, Bote Gómez (1990, p. 124) states that it is necessary to carry out an inventory of the tourist offer, by classifying its tourist resources identified using a methodology. Therefore, in a simplified way, Wahab (1991, p. 131) states that the "tourist offer can be natural or artificial and, generally speaking, encompasses everything that the destination has to offer."

As such, Bahl (1994, p. 35) considers that "the tourist offer is the set of goods and services derived from the structure of attractions, public, general and tourist utility of a locality which, when combined in different ways, make it possible to create tourist products."

In short, the tourist offer is made up of a set of attractions that are the responsibility of the public or private sector, which refer to a joint action capable of forming and maintaining a global product, building an organized offer that is integrated into the local socio-economic context.

1.4.3 The tourism product

The tourism product is made up of a "set of goods and services, united by relationships of interaction and interdependence that make it extremely complex" (RUSCHMANN, 1995, p. 11). Its singularities distinguish it from industrialized goods and commerce, as well as from other types of services. "One of the most striking features is that it is an intangible product, the residue of which, after use, is an experiential experience" (MOTA, 2001, p. 127). Because it is an abstract consumer good, it cannot, unlike tangible goods, be evaluated according to its size, weight, shape or color. It is represented to potential consumers through descriptions and images that induce the customer to buy products capable of satisfying their needs.

According to Castelli (1996, p. 60), the tourist product is made up of goods and services and adds that "tourist goods are all those that provide man with the satisfaction of physiological (rest, leisure), spiritual (pilgrimages), cultural (studies and research) and moral (changes in behavior) requirements." Mota (2001, p. 128) argues that:

> Almost all tourist goods, unlike the vast majority of goods (articles), cannot be transformed, transferred or sold. These goods can be tangible (seas, beaches, natural parks, etc.), intangible (art, folklore, prestige, image), free (air, climate) and appropriable (art collections, sports fields, etc.) (MOTA, 2001, p. 128).

> According to Tabares (1991), in a broad sense, a product is everything that contributes to consumer satisfaction. The author emphasizes that what is sold are the satisfactions that are expected to be obtained by consuming a product. According to Tabares (1991, p. 13), within tourism, the tourism product is made

up of:

> a) Attractions: these can be defined as the elements of the tourism product that determine the choice of one tourist destination over another. They refer to the natural, artificial and human environment; b) Facilities: these are the elements that contribute indirectly to the flow of tourists. The lack of them can prevent tourists from visiting the attractions. The lack of food, accommodation or transport facilities, for example, is an obvious obstacle to the realization of tourism; c) Accessibility: access is part of the infrastructure of the destination and relates to the routes and means of transport available for tourists to travel to their chosen destination (land, air or sea access routes). (TABARES, 1991, p. 13).

Tourism products have very specific characteristics that individualize them, completely differentiating them from industrial or commercial products. According to Vaz (1999), the tourist product is not a mobile good, it is a static product, impossible to stock, which depends heavily on seasonal factors and diversification, as it is an immaterial product that is subject to instability in various sectors. In other words, according to Bahl (1994, p. 55) "the tourist product can be a specific tourist destination, or an aggregate of tourist destinations, presented in the form of itineraries or tourist packages. "

In light of the above, it can be summed up that the tourism product has its own characteristics and that its intangibility makes it a unique element, made possible by the tourist trip. It should be emphasized that the attributes of services can constitute a differential in the tourist product offered, generating a competitive advantage.

It can therefore be concluded, based on the words of Vaz (1999), that the tourist product is a set of benefits that the consumer seeks in a given location and which are enjoyed with the structural support of a complex of services offered by various organizations.

1.4.4 Infrastructure and superstructure

In order for the tourism system to function, tourist attractions and facilities need to be added to the infrastructure. In the modern economy, infrastructure, according to Boullón (1997), means all the goods and services that a country relies on to support its social and productive structures. Infrastructure includes education, health services, housing, transportation, communications and energy.

Tourism infrastructure, according to Barretto (1991, p. 52), refers to the "set of goods and services that are available to tourists as an integral, fundamental or accessory part of the tourism phenomenon." Therefore, the infrastructure of a tourist center includes access infrastructure, basic urban infrastructure, tourist facilities and services, support facilities and services and tourist resources. The sum and interrelationship of these will be the city's infrastructure for tourism.

For the tourism system to function properly, in addition to the infrastructure, it requires the help of a higher subsystem that regulates the whole system, which is called the superstructure. According to Boullón (1997, p. 50), the superstructure comprises all the specialized bodies, both public and private, in charge of optimizing and modifying, when necessary, the functioning of each of the system's component parts, as well as harmonizing their relationships to facilitate the production and sale of the services that make up the tourism product. According to Beni (1998, p. 97), it comprises "the official tourism policy and its political-administrative organization, which manifests itself in the set of organizational and production measures of official bodies and institutions, and government strategies that interfere in the sector."

It can therefore be considered that the existence of a powerful and, above all, efficient superstructure is much more important than one might think, because for tourism to develop satisfactorily, the sector must maintain a direct relationship between the quality of attractions and development policies. Beni (1998) states that it is necessary to draw up guidelines for a national economic policy for the development of tourism. From an economic point of view, tourism is a multi-sectoral activity that requires coordination and planning for its development, which can only be promoted by public authorities. Another characteristic is its significant social and cultural implications, which cannot and should not only be of interest to the entrepreneur, but above all to the government, which represents the guarantee of the interests of the community.

1.4.5 Territorial articulation of the tourism system

In its most apparent manifestation, which is that of the major tourist resorts, the territorial articulation of tourism seems to have deep connections with that produced when other productive or reproductive activities are developed, such as industry, or especially construction. According to Vera (1998, p. 199),

> It should be borne in mind, in any case, that a space is not ontologically touristy unless it possesses or is the object of tourist interest. Ultimately, its condition is given by the development of a particular practice. It is the tourist's perception - rooted in their own culture and rooted in another territory - that validates its resources, its people and its elements - be they coastal, rural, natural or urban landscapes - for tourism (VERA, 1998, p. 199).

Tourist practices presuppose a system of activities, signs and conventions that identify them in contrast to non-tourist practices. Since tourists project unique expectations of use and consumption onto the places where they enjoy themselves, the condition of a tourist destination is identified by its capacity to satisfy these expectations, to allow them to develop behaviors that are different from their usual ones and to satisfy their specific needs, and by its capacity to facilitate leisure activities and/or discover new ways of living. On the basis of this uniqueness, it can be admitted that its functionality, *habitat* and dynamics are substantially different from other forms of settlement.

According to Coriolano (2006),

> The territory, represented by the places of origin and arrival of tourists; nature, meaning most of the attractions; urban-regional facilities, implying an infrastructure; people, identified as tourists and residents, are all the object of the study of geographical science and, consequently, tourism. (CORIOLANO, 2006, p. 370).

He also comments that: "The incorporation of a territory into the collective tourist imagination takes place through the conversion of its resources into products, the creation of specifically tourist attractions and services, and their promotion and commercialization." (Vera, 1998, p. 197). The way in which the tourist area is adapted, and ultimately its territorial implementation, is the result of the interaction between all the agents involved in its production, the fruit of the relationships established between landowners, urban developers, tourist entrepreneurs, direct consumers, consumer intermediaries and public agents at the most different levels. (SANCHEZ[19] , 1985 *apud* VERA, 1998, p. 197). In this way, we can see that the fundamental characteristic of this system is its difference from other economic activities, in which the destination space plays a fundamental role for all the agents.

From a territorial point of view, tourism is a great consumer of spaces, and it is also a producer and transformer of the first magnitude. For Cara (2001, p. 86) "tourism is a consumer of territory, not just a user of land, but also of the social conditions that give it meaning, and it is on this organized and meaningful territory that tourism has an impact." It is an activity that creates images and representations that impregnate the land with the agents and society that are the users of that territory.

Cara (2001) also states that:

> Local society produces images that identify it, that guide its development, that define it. Tourism itself becomes individualized, generating its own system of relationships, a field that gradually becomes independent, specifies itself and materializes in a particular society, which we can call artificial, sometimes seasonal and sometimes weekend. (CARA, 2001, p. 88-89).

Therefore, tourism is also a way of conditioning the use and, ultimately, the appropriation of a space in certain situations, and in others, of incorporating it into a space.

Rodrigues (1997) proposes a systematization of the spatial and territorial dimensions of tourism based on three organizing elements: the poles made up of the areas that send tourists, the poles defined by the receiving areas and the lines connecting these poles,

19 SANCHEZ, J. E. Por una geografia del turismo de litoral: Una aproximaciòn metodològica. **Estùdios Territoriales**, 17, p. 101-102, 1985.

through which tourists and information circulate. It is in the receiving areas that the tourism territory or tourist site par excellence can be identified.

According to Fratucci (2000), it is in the tourist place that the phenomenon materializes and superimposes its fixed forms: tourist attractions, tourist equipment and services (accommodation, food services, receptive agents, tourist guides, places and facilities for entertainment, among others) and support infrastructure (communication services, transport, security, among others). It is the place where the tourist product is produced and consumed, which, due to the peculiarities of this activity, sometimes take place simultaneously.

Therefore, the tourist site ends up bringing together space and territory. As a socio-spatial practice, tourism appropriates certain spaces, transforms them and, as a result, produces flexible and discontinuous territories and territorialities (SOUZA, 1995) and "touristifies" places. In the words of Nicolàs:

> [...] tourism creates, transforms (sic), and even (sic) differentially values spaces that could (sic) have no 'value' in the context of production logic: suddenly the pasture land can be transformed into a camping park, or the semi-derelict house of the deceased grandfather into a guest house (NICOLÀS, 1996, p. 49).

It is pertinent to complement these statements, because sometimes tourism also ends up being incorporated into a space (BAHL, 2007)[20] , whether by the public or private sector, using it as an economic resource or attraction.

Knafou (2001), in his analysis of the relationship between tourism and territory, also points out that tourists are at the origin of the phenomenon, and that it is they who define and choose tourist sites. His proposal is that we shouldn't lose sight of the fact that the subject of the phenomenon, responsible for its existence, is man, in the form of the tourist. The author suggests the possibility of three types of relationship between tourism and territory: a) there can be territories without tourism; b) there can also be tourism without territory and; c) finally, there can be tourist territories, the latter understood as "territories invented and produced by tourists, more or less taken over by tour operators and planners" (KNAFOU, 2001, p. 72). It is in the latter that conflicts of territoriality arise between tourists, nomads and sedentary hosts:

> [...] there are different types of territoriality that clash in tourist places: the sedentary territoriality of those who live there frequently, and the nomadic territoriality of those who only pass through, but who have no less need to appropriate, even fleetingly, the territories they frequent (KNAFOU, 2001, p. 64).

For Falcâo (1996), tourism, described as a new form of mass consumption, is developing within the emerging economy of invisible exchanges on a national and international scale.

20 BAHL, M. Aulas da disciplina de produçâo turistica e efeitos sociais, economicos e culturais no espaço urbano e regional. Curitiba: Postgraduate Program in Geography - UFPR, 2008.

This modality expands with the production of goods (infrastructure, buildings, food and various products) and services (transportation, accommodation, food, among others), which are integrated for final consumption. This set of goods and services offers the consumer market the 'conditions of accessibility' to a given place. "Space, in the dimension of place, takes on the character of a consumer object and, as such, is (re)produced and marketed" (FALCÂO, 1996, p. 65). We can therefore conclude that it is the territory plus a certain (symbolic) value and territorialized infrastructures (lodging facilities, theme parks, airports, etc.) that becomes a product and as such is sold and consumed.

"The analysis of tourism and its territorial dimension is one of the means of (re)thinking this multiform and inapprehensible phenomenon, doubly marked by the weight of economic interests." (KNAFOU, 2001, p. 72-73). Global transformations and the speed of change increase the possibility of aggressive impacts on socio-territorial structures and it is in this sense that constant observation is required of processes that should be planned and managed consciously and with the aim of improving existing structures.

1.5 THE TOURIST AREA

The tourist area is the consequence of the presence and territorial distribution of tourist attractions, which are the raw material of tourism. "Elements such as tourist heritage, the resort or attraction and the infrastructure are enough to define the tourist space of any country." (BOULLÓN, 1997, p. 65).

In Brazil, we find the approach proposed by Santos (1991, p. 65), centered on space, according to which "social space, like all social reality, is defined methodologically and theoretically by three general concepts: form, structure and function". Santos (1994) later added another category for analyzing space: process.

According to Bonfim (2007), analyzing these categories makes it possible to understand the dynamics of the tourist area and the social processes that produced it. According to Santos (1991), these categories form the basis of the method for interpreting the tourist area.

For Santos (1997, p. 50), form "is the visible space of a thing. It also refers to the orderly arrangement of objects, to a pattern. Taken in isolation, we have a mere description of phenomena or one of their aspects at a given moment in time". This would mean that form is the visible aspect of tourist space, that which exists in material form, which is observable. Is it the landscape? Santos (1991, p. 72) states that:

> [...] Landscape is different from space. The former is the materialization of an instant of society. [...] Space results from the marriage of society and landscape. Space contains movement. Landscape and space are therefore a dialectical pair. They complement and oppose each other [...] (SANTOS, 1997, p. 50).

To make his point clearer, Santos (1991) uses the example of the city of Salvador on a given day and at a given time. There is a certain distribution of people and production over the territory. A few hours later, the distribution is no longer the same. Everyone produces: the worker, the artist, the salesman, etc., everyone who consumes it. The way in which production takes place between men and the exchange between men gives the appearance of a landscape. Accumulated work and life happen together. Accumulated work is the landscape. Space is the combination of accumulated work and living work, of geographical forms and social context.

Space is therefore "the result of the sum and synthesis, always redone, of landscape with society through spatiality. Landscape has permanence and spatiality is a moment". (SANTOS, 1991, p. 73).

The approach of this category has a special connotation for tourism studies. However, Bonfim (2007, p. 50) argues that "it cannot be dissociated from the other categories, otherwise it is reduced to a mere description of the landscape or to empiricism."

Santos (1997, p. 51) also explains that:

> [...] form can be imperfectly defined as a technical structure or object responsible for carrying out a certain function. Forms are governed by the present, and although it is customary to ignore their past, it remains an integral part of them. They emerged endowed with certain contours and purposes-functions (SANTOS, 1997, p. 51).

Bonfim (2007, p. 50) complements this idea by arguing that:

> Structure refers to the social and economic nature of a society at a given moment in time. Without structure, it is not possible to understand the historical nature of space, because it includes form and function. It expresses spatial functionality (BONFIM, 2007, p. 50).

According to Corrêa (2000), structure is the social matrix in which forms and functions are created and justified. This category is highly valued in the planning and management of tourist territories, taking on an economic focus.

Santos (1997, p. 50) describes function as "a task or activity expected of a form, person, institution or thing (...). It is the elementary activity of the form. The latter may or may not encompass more than one function". Function implies the ability to distinguish, for analysis purposes, certain elements of the tourist space, such as: supply, demand, transport, infrastructure, services, management and *marketing,* which occur simultaneously or at a given historical time.

Form can be expressed as revealed structure. Santos (1997, p. 51) refers to it as:

> Being the most visible, it is apparently, and to a certain extent, easier to analyze than the structure. The forms or artifacts of a landscape are the result of past processes in the underlying structure. However, divorced from structure, form

> will lead to a false analysis: in fact, similar forms have resulted from extremely diverse past and present situations [...] (SANTOS, 1997, p. 51).

Santos (1997, p. 50) defines a process as "a continuous action, developing towards some result, implying concepts of time (continuity) and change". The process can be considered as an action that is carried out, as a rule, continuously, aiming at a result that implies time and change. The process seeks to analyze the dynamics of space, the transformations that tourism places on territories, imposing leisure representations on pre-existing ones. "The dynamism of space can present periods of stability, of small changes and times when complete transformations take place, producing new spaces. The process is the genesis of tourist territories" (BONFIM, 2007, p. 51).

This framework presents the methodological difficulty of how to differentiate the socio-spatial transformations resulting from tourism from those resulting from other social practices.

"When technicians work on determining the tourist area, what they do is delimit a flat surface on the map, which is the best way of representing the area that interests physical planners." (BOULLÓN, 1997, p. 65). It can be seen from Boullón's statement that the tourist area cannot be delimited using common regionalization techniques, because according to these techniques it would have to cover the entire surface of the municipality, state or country to be studied. For Boullón (1997), this would be a mistake. Boullón claims that there is no such thing as a tourist region, and that through his studies of regions he developed the theory of tourist space.

"The best way to determine a tourist area is to use the empirical method, because you can observe the territorial distribution of tourist attractions and the land, in order to detect the groupings and concentrations that appear" (BOULLÓN, 1997, p. 66). Thus, by means of a systematic procedure and with the application of specific methodologies, Boullón (1997, p. 66) indicates which are the components of tourist space: "zone; area; complex; center; unit; nucleus; set; corridor; transfer corridor; stay corridor."

It can therefore be concluded that the theory of space is a useful tool to guide the analysis and diagnosis of the sector, using a procedure that simplifies intellectual work by conducting it in an orderly manner. Subsequently, and according to the potential of each element and the demand projections, first of all, growth limits should be established for all the elements of the tourist area analyzed and then the categories and types of tourist plants to be built at each stage of their evolution should be calculated with a satisfactory level of approximation. In the resolution part of a plan, after knowing what the elements that make up the tourist area are and how they work, it is necessary to identify projects, designed as a way of improving the individual performance of each element in a coordinated way, imagining that each partial success will result in a joint benefit.

1.5.1 Tourism in coastal areas: from growth to the restructuring of space

Vera (1998) argues that tourism in coastal areas is the most common and differential form of tourism development, and is the one that generates most of the international tourist movements, since the coast is the main destination in many countries.

The high consumption of land associated with tourism on the coast, with all its diversity of implementation models and extraordinary capacity for transforming territorial, social and economic structures, makes tourism in coastal areas a type of activity that requires special attention, especially in the current phase of restructuring, in the face of the problems and contradictions arising from the accelerated growth of recent decades and the necessary adjustment of tourist destinations that must be molded to the new motivations of demand.

Tourism is one of the newest forms of the accumulation process, which has been producing new geographical configurations and materializing space in a contradictory way, through the actions of the state, companies, residents and tourists. Understanding this dynamic means understanding the productive relations of space and the exercise of power by the state, business and working classes in movement and conflict. According to Coriolano (2006), in order to reproduce itself, tourism follows the logic of capital, when a few appropriate the spaces and the resources they contain, presenting them as attractions transformed into commodities.

This analysis tries to get away from the structuralist approach which considers space as a rigid, homogeneous, isotropic, neutral object, delimited by peripheral lines, localized, formal, a *priori of* man, the stage for events, and to admit it as a historical and dialectical process resulting from social relations of production. Thus, geographical space is neither a support nor a reflection of society's actions, but a social product. Santos (1979, p. 18) shows that:

> Space reproduces the social totality insofar as these transformations are determined by social, economic and political needs. Space reproduces itself within the totality when it evolves as a result of the mode of production and its successive and contradictory movements (SANTOS, 1979, p. 18).

Based on this understanding, physical space itself is now considered from the point of view of its organization and social meaning, creating spatialities, i.e. spaces produced through relations of force and power that are established in a contradictory way, transforming so-called first nature into second nature. According to Coriolano (2006, p. 369): "Space is thus the main category of geographical analysis and it contains a series of other categories and supporting concepts, such as: territory, place, region and landscape, among others." While territory is "the historical result of society's relationship with space, which can only be unveiled by studying its genesis and development" (MORAES, 2002, p. 63). It is the place, the space of resistance, where everyday struggles are waged, the exploitation of labor forces, the flow of surplus value and the productive restructuring of capitalist accumulation. Struggles that previously seemed to be only between social classes are expanding and reaching places. Not only are classes fighting for their antagonistic interests, but spaces and places are becoming competitive and threatening, resulting in what Santos (1994) called a "war of places". Spaces are being produced differently as a form of subsumption to capital.

Tourism is currently one of the triggers of this spatialization, acting by deterritorializing/reterritorializing and producing new geographical configurations. Thus, coastal regions, originally occupied by indigenous people, fishermen and traditional communities (the so-called "peoples of the sea"), are expropriated to make way for second homes, large *resorts,* hotel chains, restaurants and other tourist facilities, such as theme parks, for example. Coriolano (2006) argues that in this spatial production, it is necessary to consider the struggle of the different local actors: the native users of the coastal space who are trying to defend their properties, or the goods they use, against the interests of businessmen, real estate agents and the state itself, who are interested in the exchange value of the space, since they have transformed it into a commodity.

Tourism is materialized in the logic of the historical and geographical differentiation of places and regions. It is as pertinent to the local as it is to the global, because it dominates historical social relations as a result of the changes and restructuring of spaces, taking advantage of local resources. It transfers the value of cultural heritage, cities, places and the local population to tourists, as an object of gaze, pleasure and desire. Coriolano (2006) argues that tourism and the consumption of spaces produce various structural forms of landscapes and businesses. At a time when historical processes would be inoperative, the technique used provides agility and the capacity for territorial organization. There is a spatial demand for each type of tourism. Mountain, mountain and coastal areas, urban and rural areas, metropolises and historic cities, even deserts and enclaves are used.

The people living in the coastal areas are fighting a resistance struggle to remain there. The advance of real estate speculation is stimulated by the investments and infrastructure put in place by the state. Many residents of areas near the sea have been displaced to less favored areas by the growing construction of second homes near the sea. According to Coriolano (2006), this was a trajectory of compulsory displacement, followed by conflicts between coastal communities and real estate speculators. With each need for modernization imposed by productive restructuring, whether influenced by the state or not, the seafront along the entire coast has undergone new uses and appropriations.

The conclusion is that the production of tourist space is determined by socio-spatial and power relations. That tourism is a social practice of appropriation and domination of many territories, especially in tropical countries. That the right to the city, to leisure, to tourism and to reclaimed space continues to be the subject of much study.

1.6 MASS TOURISM

In developed countries, tourism and leisure in general are more accessible to the population as a whole. According to Krippendorf (1989), it is common for workers in developed countries to spend their vacations with their families in summer resorts in the countryside, in the mountains or on the beaches. It can even be in water resorts or *spas*. Tourism for workers in general, young people and children, the elderly, the physically disabled, naturalists, ethnic, cultural and religious minorities, is becoming a reality every day, but the biggest problem is that these tourists are largely seasonal, making it difficult for tourist destinations to function properly outside the vacation season.

It is important to realize that the maintenance of regular tourist flows through a given territory depends, in addition to infrastructure and specific investments, on the relationship that the tourist will have with the native population. Respect must be maintained on both sides and cultural, ethnic, economic or social differences must learn to live together.

Mass tourism is a consequence of contemporary society, for Krippendorf (1989, p. 23), "mass tourism constitutes one of the most striking and heaviest in consequences and least curbable forms of the shock of leisure." But society has learned to live with it. It is believed that excesses can be contained, that developments can be predicted and weaknesses remedied. In fact, looking at the current situation, it is difficult to do anything but contemplate the effects that mass tourism will still have on the ecological, psychological and socio-economic levels. Unless people's mentality changes, the system is altered and new overall conditions are developed.

It is important to emphasize that mass tourism, in its deepest aspects, is not a separate world that obeys its own laws. Its aspects are both a consequence and a component of the social and industrial system, the organization of human beings and modern civilization.

What is noticeable in mass tourism is the predominance of economic interest, the more people, the more foreign exchange, the more return. Is that so? What kind of return do the locals expect? Just the economic return? That's not enough. The negative impacts of mass tourism are often far greater than the financial return achieved. But what are these impacts? Degradation and/or destruction of the natural environment, perhaps the loss of local identity, the loss of local people's prospects of another form of tourism, or of producing financial rewards, social and political problems, etc. Looking at it from this angle, it's important to stress the need to propose alternative forms of tourism in the receiving areas. Mass tourism is generally linked to certain times of the year and the consequent seasonality, which is so difficult to break in many places.

1.7 SECOND HOME TOURISM

Assuming that tourism is a complex activity with multiple economic, social, political and cultural relationships, it is necessary in this context to analyze the socio-spatial repercussions of second home tourism. Its expansion on a global scale has led to important changes in the places where it is developed. In Brazil, the first studies on this phenomenon date back to the late 1980s (TULIK, 1995). However, several questions remain unanswered.

Second homes are a type of accommodation linked to weekend and vacation tourism. Despite its historical and intense expansion on a global scale, second homes are still a little-studied and little-known phenomenon that suffers from a lack of a solid base of theoretical reflections and empirical studies of its various socio-spatial repercussions in different parts of the world.

In most tourism studies, second homes are a one-off topic, often mentioned as one of the types of accommodation in the areas in question. Case studies that focus discussions on this topic are still scarce and, for the most part, are located in the main countries of Europe and North America. Pearce (1991, p. 116) points out that in these studies "the main objective is to analyze the distribution of second homes and not to analyze the spatial structure of tourism in general."

Soneiro (1991, p. 126) adds that these emerging studies on second homes are "[...] largely idiographic and descriptive in nature, particularly oriented towards the study of distribution and spatial impacts".

Hά, therefore, the need to deepen specific methodologies for the study of second homes. A worldwide phenomenon of recent analysis, second homes began to appear in international literature in studies from the mid-1970s, according to Pearce (1991). According to Becker (1995, p. 10):

> In Brazil, the phenomenon of second homes emerged in the 1950s under the aegis of "national developmentalism", which was responsible for the establishment of the automobile industry, the rise of the road as the main transportation matrix and the emergence of new middle and urban social strata that gradually began to incorporate the ideology of tourism and leisure among their socio-cultural values. Summertime or weekend breaks became a social value, the satisfaction of which would lead tourism, in an often predatory and disorderly manner, to regions accessible to large urban centers in the Center-South, and with valued environmental attributes (coastal and/or mountainous areas). (BECKER, 1995, p. 10).

According to Assis (2003), the growth of second homes in Brazil and the provision of statistical data by the IBGE over the last three decades have already resulted in some specific analyses of the phenomenon in the country, such as the pioneering study by Odette Seabre (1979) on the socio-spatial effects of second homes (land prices and forms of occupation) on the coast of Santos-SP; Mauren Roque's (1990) analysis of the social inequality that the phenomenon of second homes entails as a manifestation of the differences in work and leisure opportunities in the seaside resort of Guarujà-SP; and Olga

Tulik's (1995) work evaluating the size and distribution of second homes in the state of São Paulo, highlighting the most significant source and receptor areas. These analyses, produced as dissertations and theses at the University of Sâo Paulo, each present proposals for theoretical reflections and empirical analyses which, together with the international literature consulted, serve as references for reflection on the phenomenon of second homes.

That said, despite the IBGE's technical definition, the secondary residence or second home is a broad and complex concept which, due to the profusion of restrictive terms - beach house, summer house, country house, seasonal house, vacation house etc. - still lacks a terminological consensus.

Based on a limited foreign bibliography exclusively on second homes, Tulik (1995) sets herself the challenge of analyzing the conceptual aspects of this type of tourist accommodation in relation to the Brazilian reality, specifically that of São Paulo, her object of study.

Tulik (1995, p. 21) defines a secondary residence as "[...] a private tourist accommodation used temporarily during leisure time by people who have their permanent home elsewhere".

This concept, which is much broader than that of the IBGE, leads to a discussion of the main links that engender it and make second homes a complex phenomenon.

Firstly, the second home is considered to be private tourist accommodation, i.e. privately owned. This question of ownership brings us back to the basic characteristic of the second home - income as the differentiating factor that defines who can own, in addition to their permanent home (first home), another home for weekend and vacation leisure.

The second home presupposes the availability of a surplus income, as it implies costs for buying the land, building the property (when you don't buy the property), taxes, maintenance and means of transportation for commuting (usually a private car). Assis (2003) argues that these factors make second homes an elitist type of tourist accommodation, a symbol of social *status,* characteristic of the upper and, for the most part, middle social classes. For the latter, the lack of financial availability and free time to enjoy vacations with the family on long trips around the world makes second homes an important leisure alternative, due to the savings in time (work) and, above all, money.

In addition to accommodation as a tourist leisure alternative, second homes are an investment option. According to Soneiro (1991, p. 151), second homes "[...] are, above all, a profitable investment, as evidenced by the following fundamental factors: their low frequency rate, since they remain empty for most of the year, and the short time they generally belong to the same owner...". Despite the high maintenance costs and, in general, the low frequency rate of secondary residences, ownership of these homes represents an "investment in land", as reserves of value for the future (TULIK, 1995). Assis (2003, p. 112) explains that "[...]

investment in land is the safe way to preserve accumulated capital (especially in an inflationary economy) and consequently to expand capital through the appropriation of financial income."

However, Tulik (1995, p. 24) warns that second homes "[...] are an investment that does not offer liquidity or immediate returns, as the sale is subject to the laws of supply and demand and these properties are not always rented out, remaining empty for most of the year".

The question of ownership of second homes also comes up against a very common practice related to the use of these tourist accommodations: many owners, in order to obtain a surplus income, rent out their second homes to other tourists, rent them out or lend them to relatives and friends. This fact, which is not captured by the IBGE's technical concept - despite the fact that second homes are part of private households - makes the definition of the term complex.

According to Tulik (1995):

> [...] the secondary residence, as a private property (since there are no public secondary residences, at least in Brazil), is a type of tourist accommodation whose operational concept should not be linked to the fact that it is owned, rented, leased or borrowed (TULIK, 1995, p. 18).

Another issue that the concept of second home raises is temporality. As a home for occasional use at weekends or during vacation periods for recreational and leisure purposes, the second home undoubtedly presupposes the existence by the owner of another home for permanent use (also known as the main or first home) where they live for most of the year. According to Assis (2003), while:

> [...] the first home responds to the most basic need to live, to dwell, to have shelter, despite the different ways of satisfying these needs, the second home escapes any understanding of this nature. It is a second home (ASSIS, 2003, p. 113).

No matter how long they stay in the second home, the factors of free time, non-work, surplus income and distance from the main home define the temporary use of these buildings.

> Some authors note that use can be repeated, but not consecutively for a period of more than a year, which establishes the territorial link and a certain parallelism with the accepted definition of tourist, a fact that reinforces the notion of secondary residence as tourist accommodation (TULIK, 1995, p. 21).

The third issue that permeates the concept of second homes is their purpose. A hallmark of "post-industrial" society, the right to free time is one of the great social conquests of the world of work, and second homes are one of the ways of enjoying it. The pursuit of recreation and the enjoyment of free time is undoubtedly the main purpose for which second homes are built.

With the industrialization and metropolization of cities, it became increasingly necessary for urban people to move away from the overcrowded central areas towards the metropolitan outskirts in search of a reunion with nature. This was a way of relieving their daily stresses and renewing their energy.

The city, which was once the center of attraction for rural dwellers and men in search of work in industry, now - despite still concentrating various functions - "expels" its residents because it doesn't offer them the necessary conditions for using their free time in contact with nature (ASSIS, 2003).

Due to their proximity to the central areas, the peripheral metropolitan regions became the main targets of real estate speculators and tourism *marketing* strategies that sought to enhance the natural and cultural attributes of these spaces, offering them to specific social segments that had surplus income to purchase a secondary residence.

At this crossroads that leads to the "natural-urban", the second home is one of the maximum expressions, since its conception is associated with the commercial capitalization of the natural attributes of landscapes - the sea, the countryside, the mountains, etc. - and the provision of an infrastructure of urban facilities and services that facilitate greater use of leisure time.

The second residence turns the "weekend" into a socio-cultural factor characteristic of contemporary society. The reduction in the working day, the degradation of the urban environment and the advent of the private car have all contributed to the emergence of the weekend as the main period of free time, especially via the second home (ASSIS, 2003). For Krippendorf (1989), the weekend represents a space of freedom in which we can see that free time is a privileged opportunity in which man can find himself, while at the same time allowing him to recover from the adverse situations of the social environment and work.

Generally, tourists begin their migrations to their second homes on Friday evenings, after their weekly working day or on Saturday mornings, thus taking advantage of Saturday and Sunday, and generally returning to their permanent homes in the afternoon or evening of Sunday to resume their work on Monday morning (ASSIS, 2003).

As a recent phenomenon linked to weekend commuter tourism and annual vacations, Tulik points out that:

> The location of secondary residences is defined by the time-cost-distance relationship. These factors can contribute to the appreciation of natural and cultural resources closer to the centers of origin, which are preferred to others of higher quality but further away (TULIK, 1995, p. 202).

Based on Tulik's considerations on the time-cost-distance relationship, it can be inferred that this relationship can, in some cases, define the location of second homes in urban or rural areas and influence the concentration of these buildings in certain areas. Also, the availability of income and free time influences the choice of location of the second home, since the further away from the permanent residence, the greater the time and cost of access to this home for occasional use. In addition, owning a second home in areas very far from the main residence makes it impossible to visit it regularly at weekends, leading the owner to use it mainly during the vacation season, when they can spend more time there. However, nowadays, with the improvement in the quality of access roads and the technological advances in means of transportation (especially private cars), the distance factor can, in some cases, become secondary or irrelevant when choosing the location of the secondary residence.

Pearce (1991, p. 16) points out that:

> The main factors in the distribution of second homes seem to be: the distance from large population centers; the quality or attributes of the landscape; the presence of the sea, rivers or lakes; the presence of other recreational resources; the availability of land; the climates of the sending and receiving regions (PEARCE, 1991, p. 16).

It can be seen from Pearce's comments that second home areas tend to move away from urban centers as they expand and urbanize, migrating from the metropolitan periphery to rural areas, and that the location of these homes can also be influenced by cultural factors.

The different realities generated by second homes confirm their complexity and require scholars of these tourist accommodations to be cautious when interpreting and generalizing their conclusions. By articulating various spheres of socio-spatial organization, second homes dynamize the spaces where they are installed, triggering positive and negative effects in the most different places and realities.

As examples of this complexity, Tulik (1995, p. 138), recognizes the positive effects of second homes in some São Paulo municipalities, stating that "the increase in occasional-use homes tends to generate a lot of revenue for the municipalities, while also improving their social indicators in terms of the number of telephones and residential electricity consumption in general".

In contrast, Seabra[21] (1979 *apud* ASSIS, 2003) highlights the negative effects of second homes, such as apartments, on the coast of Santos which, following the capitalist logic of commodity production, led the city to lose its beach conditions and emerge as a center of commerce and services.

In an informal conversation with tourists and residents of the coast of Paraná, it emerged that

---

21 SEABRA, O. C. de L. **A muralha que cerca o mar: uma modalidade** de **uso do solo urbano**. 122 f. Dissertation (Master's Degree in Geography) Faculty of Philosophy, Letters and Human Sciences, University of São Paulo. Sâo Paulo, 1979.

their opinion was practically the same, i.e. that the unbridled capitalization of the coast for the construction of second homes has led to the degradation of the natural landscape and the cooling of tourism itself. This, consequently, also leads to a loss of sustainability for the "native" population, as many of them have their jobs tied to tourism.

These spatial repercussions - positive and negative - are generally associated with the expansion of urbanization and the process of real estate speculation in the receiving areas, the diversification of the job market and its effects on income patterns, the proliferation of "seasonal residents" and the loss of tourist attractiveness.

Among the positive effects are the changes that have taken place in the world of work, which generally lead to an increase in trade and service activities, creating more job opportunities and higher incomes.

The most common negative repercussions, according to Barros (1998), refer to the fact that second homes are usually..:

> [...] with their buildings and infrastructures, they displace the previous traditional agricultural and fishing land use and cause turbulence at cultural and social levels (employment profiles, changes in lifestyles and horizons, etc.) and environmental levels (disorganization of drainage and geomorphological changes, destruction of the quality of fresh and ocean waters, etc.) (BARROS, 1998, p. 28).

In Guaratuba, for example, it can be considered that the impacts of mass visitation and the disorganized growth of second homes on the coastal plain have caused socio-environmental implications by overloading both natural ecosystems and public services such as water supply, sewage disposal and solid waste collection during the high season.

Furthermore, according to Assis (2003), the frequent refuge in secondary residences establishes identities and "territorialities" that often trigger spatial repercussions and socio-cultural conflicts between the "vacationer" and "native" populations.

Analyzing all this, it can be seen that the socio-spatial transformations triggered by second-home tourism open up a wide field of study that enables and calls for a geographical analysis of the phenomenon.

1.8 tourism planning

Tourism has become one of today's most unique phenomena. In order to discover its nature, it is necessary to try to understand how its elements are connected, what the causes and effects are, the desires and realities of the people who carry it out. According to Krippendorf (1989, p. 24), "we must first master how the mechanism works before we can determine the means of controlling, modifying and perfecting it." In other words, planning tourism, because everything that is planned is more organized and brings better results.

For Ackoff[22] (1976 *apud* BARRETTO, 1991, p. 11), planning "is a process that is intended to produce one or more desired future states that should not occur unless something is done." Adequate and orderly planning is essential for the development of tourism in any location. It is important to remember that, "planning is an activity, not something static, it is a becoming, a happening of many concomitant factors that have to be coordinated in order to achieve an objective that is at another time." (BARRETTO, 1991, p. 25). As such, it is a dynamic process that is open to constant revision and course correction, and can therefore require constant rethinking, even after the objectives have been achieved.

Tourism planning can be classified into several types, according to different approaches. According to Barretto, the aspects of planning are:

> Temporal: related to time, to the deadline; geographic: whether it is worldwide, local, subdivided into rural and urban and some specific cases, maritime, fluvial, mountain, etc.; economic: linked to geographic, it presupposes a territory, associated with macro- and micro-economics; administrative; intentional and aggregative. (BARRETTO, 1991, p. 19).

A planner can't work on the basis of empiricism, "let's do it and see what comes of it", but rather an in-depth study of the whole present context, of the socio-economic situation in which the plan is inserted, as well as of the planner himself. "The planner must also know how to interpret the facts analytically and choose whether the interpretation of the data should be quantitative or qualitative." (BARRETTO, 1991, p. 14).

In a nutshell, then, planning consists of defining objectives, organizing resources, determining methods and forms of organization, establishing time measures, spatial location, etc. According to Barretto (1991, p. 14), "planning requires an understanding of the problems and a harmonious distribution of specialties; it requires knowledge."

Planning helps people or organizations to see the near or distant future, helps tasks to be carried out better and objectives to be achieved more easily. According to Petrocchi (1998, p. 20), "it orders actions and prioritizes them, allows difficulties or obstacles to be mapped out and thus alternative paths to be chosen in advance."

Tourism planning can be divided into three levels according to Barretto:

> a) First-level planning: events, excursions, trips, etc;
>
> b) Second-level planning: transforming cities into tourist hubs, activating pre-existing tourist hubs, creating tourist complexes or cities (building tourist facilities);
>
> c) Third level planning: national policies to encourage tourist activity in the country and organize it, covering the other two levels (BARRETTO, 1991, p. 63).

---

22 ACKOFF, R. **Planejamento Empresarial**. Rio de Janeiro: Livros Técnicos e Cientificos, 1976.

It's important to understand that in order to plan a tourist activity, it's necessary to know about it so that it can be exploited in the best possible way to avoid causing so many impacts on the environment and the community in which it's located. Planning tourism is not an easy activity; it is complex and depends on a qualified and capable workforce.

Seasonality will then be addressed, which in a way is often linked to local planning, as well-planned tourist sites tend to suffer less from the seasonal issue.

CHAPTER II - SEASONALITY

Seasonality, in its contextual sense, can be defined as a certain period for the occurrence of a phenomenon, i.e. "that which occurs in some periods and not in others." (MOTA, 2001, p. 98). For Ruschmann (1995), tourist seasonality is the result of the concentration of tourist activities in space and time.

According to Dencker (1998), many factors can influence tourism demand: demographic factors, sociological factors, economic factors, tourism factors and seasonality. The seasonality of tourist demand takes into account the following variables: school or workers' vacations, purchasing power, space-time concentration (MOTA, 2001).

Regardless of the variable, the occurrence of tourism seasonality, according to Mota (2001), has consequences at various levels:

> It generates unemployment, mortality in micro-enterprises, a drop in the turnover of tourist companies, changes in the management system, compromises the quality of service, modifies the promotional policy of the tourist product, changes prices, demands greater administrative flexibility, etc. (MOTA, 2001, p. 98).

Seasonality is something much more complex and its study can identify its impacts, its problems and rationalize solutions that are appropriate in each situation or receiving community.

According to Silveira (2010):

> It is common knowledge that geographical location and climate have a strong influence on tourism demand. It is because of these factors that part of Brazil's coastal territory, like many other destinations in peripheral areas, suffers the influence of seasonal tourism. A condition that leads to the creation of seasonal tourism economies, especially in the coastal areas of the south of the country (SILVEIRA, 2010, p. 1, in press).

But climatic and geographical factors should not be seen as the sole culprits of seasonality. The problem can be triggered by a number of factors arising from a variety of reasons, both at the receiving and sending ends.

2.1 DEFINITIONS OF SEASONALITY

The growing importance that tourism has assumed in global terms and particularly in the context of some Western economies that are becoming less and less competitive in terms of traditional economic activities, justifies the great attention of the public and private sectors, both in terms of obtaining a more in-depth knowledge of the tourism phenomenon and, more specifically, in terms of the constraints inherent in its affirmation and progress. In this field, one of the issues of concern to a substantial number of destinations is the economic, social and environmental sustainability of tourism growth. Closely related, another problem is the strong three-dimensional concentration (temporal, spatial and at product level) that generally persists in characterizing tourist demand.

> In particular, temporal concentration - or seasonality - is generally recognized in the literature as a predominantly economic phenomenon, since it is consensually related to the adversity of its consequences at this level, both for the destination areas and for the companies that operate in them (ALMEIDA; KASTENHOLZ, 2009, p. 2).

The negative effects of seasonality, which include, above all, production inefficiency and a substantial limitation on the sector's annual income, mean that this phenomenon is seen by the destination area management bodies as a weakness or a major problem to be solved.

Although the seasonality of tourism demand is a widely studied phenomenon, some authors characterize it as poorly understood, at least in some of its facets (BUTLER 1994; BUTLER; MAO[23] 1996 *apud* ALMEIDA; KASTENHOLZ, 2009, p. 4). As mentioned above, it can be said that seasonality is accepted as a global tourism problem. The phenomenon affects the majority of tourist destinations, but in particular those furthest from the equator, as well as the more peripheral ones in general, which is why, especially in these areas, seasonality has been one of the prime targets of public policies and private management measures aimed at reducing it and mitigating its effects (ALMEIDA; KASTENHOLZ, 2009). However, the obvious persistence of the problem leads us to believe that a substantial part of these policies and measures are not achieving their objectives, which possibly reinforces the argument that there is still little understanding of the phenomenon and, in particular, its causes. In essence, it seems to be an expressed inability to understand and get to the heart of the problem. Is it therefore advisable to explore in detail the seasonal structure of demand in the destination area, including an analysis of the profile of tourists at different times of the year (characteristics and motivations) and, above all, the associated causes, so that the most appropriate and effective strategies can be designed to deal with seasonality?

At the same time, substantial parts of the definitions of seasonality in tourism literature often focus on the nature of the phenomenon, both in terms of its causes and its consequences. In the first study published on the subject, BarOn[24] (1975 *apud* ALMEIDA; KASTENHOLZ, 2009, p. 4) "highlights the effects of seasonality on destinations and defines it as the concentration of demand that occurs each year, more or less, in the same time frame and with the same magnitude."

More recently, perhaps the most widely referenced definition is that of Butler[25] (1994 *apud*

---

23 BUTLER, R.; B. MAO. **Seasonality in Tourism:** Problems and Measurement. Quality Management in Urban Tourism. P. E. Murphy. Chichester: Wiley, p. 9-23, 1996.
24 BARON, R. R. V. **Seasonality in Tourism:** A Guide to Analysis of Seasonality. Economist Intelligence Unit Technical Paper, n. 2, 1975.
25 BUTLER, R. **Seasonality in tourism:** Issues and problems. Tourism: The State of Art. A. V. Seaton. Chichester: Wiley, p. 332-339, 1994.

ALMEIDA; KASTENHOLZ, 2009, p. 4-5), who presents seasonality as "a temporal imbalance in the tourism phenomenon, which can be expressed in terms of dimensions such as: number of visitors, visitor spending, traffic on highways and other forms of transportation, employment and tickets at attractions." Other definitions highlight the concentration of tourist flows in short periods of the year, promoting, on the one hand, peaks of activity that often constitute a heavy burden on the physical and social resources in the area of a destination and, on the other, situations of high demand that generate inefficiency in tourist activity (LAGE; MILONE, 2000).

In short, there are many agents and interest groups involved in and affecting a country's tourism and, consequently, the flow of people at different times of the year. The incorporation of perspectives, objectives and joint efforts are vital for planning and developing an increasingly less seasonal tourism.

The seasonality of tourist demand can cause inflation in the receiving community, since if demand grows and supply has reached its maximum capacity, unable to satisfy demand, prices rise. Both essential and supplementary sectors react by selling their goods and services in a seasonal market (WAHAB, 1991).

According to Lage and Milone (1998, p. 61), "the existence of seasonal tourism demand, which is short term per season, harms tourism supply, which becomes a serious problem for the development of the activity."

2.2 CAUSES OF SEASONALITY

The difficulties in identifying and structuring the causes of seasonality in greater depth seem to result, firstly, from the great complexity attributed to the phenomenon and, secondly, from the strong and intricate correlation between them. In this regard, Butler[26] (2001 *apud* ALMEIDA; KASTENHOLZ, 2009, p. 5) "argues that the phenomenon needs further research into the detailed identification of its root causes, particularly those related to tourist motivations and consumption patterns."

It can be said that the factors explaining the concentration of tourist demand vary from destination to destination, depending, among other things, on their geographical location, the different origins of tourists and the attractiveness of each destination.

The correlation structure between the factors also has different effects on the different destinations. There are, however, common factors that can justify seasonality in most destination areas. Mota (2001) suggests seasonality and the cause and effect relationship as illustrated in figure 2, and some authors suggest grouping them into categories, as illustrated in table 1 cited by Almeida and Kastenholz.

---

26 BUTLER, R. **Seasonality in Tourism**: Issues and Implications. Seasonality in Tourism. T. Baum and S. Lundtorp. Oxford: Pergamont, p. 5-21,2001.

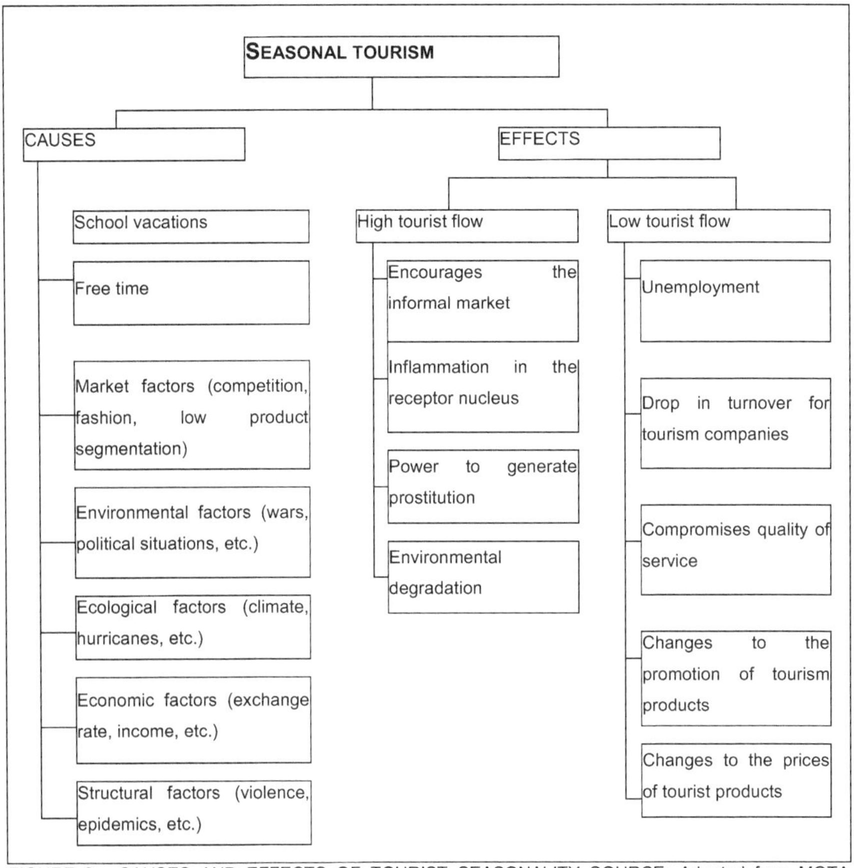

FIGURE 2 - CAUSES AND EFFECTS OF TOURIST SEASONALITY SOURCE: Adapted from MOTA (2001, p. 99).

The figure above describes in a basic way what comes closest to the Brazilian reality in terms of the causes and effects of seasonality. Knowing the causes of seasonality helps planners to focus the development of tourism in their territory, seeking efforts to reduce seasonal fluctuations as much as possible. Knowing the effects of this seasonal season makes it interesting not only for the planner, but also for all the companies and organizations involved in the tourism process to concentrate their efforts on trying to make the distribution of tourist demand more homogeneous and regular throughout the year.

The table below categorizes the causes according to authors who view and study seasonality at an international level, placing the causes mentioned in figure 2 into the categories explained below.

Author	**Categories of Causes of Seasonality**

BarOn (1975)	Natural seasonality, institutional seasonality, the effects of different calendars, sociological and economic causes.
Hartmann (1986)	Natural seasonality, institutional seasonality.
Butler (1994)	Natural seasonality, institutional seasonality, social pressure and fashion, sports seasons, tradition and inertia.
Butler and Mâo (1996)	Physical and socio-cultural factors in the tourist areas of origin and destination.
Frechtling (2001)	Climate, social customs/holidays, professional customs, effects of different calendars.
Baum and Hagen (1999)	Like Frechtling (2001), but adding the constraints of supply.

TABLE 1 - CLASSIFICATION OF THE CAUSES OF SEASONALITY IN TOURISM

SOURCE: Koenig-Lewis and Bischoff[27] (2005 *apud* ALMEIDA;KASTENHOLZ, 2009, p. 5-6).

From the table above, two categories seem to meet with greater consensus: natural and institutional seasonality. There are, however, other important causes of seasonality which are related to motivational aspects and which could be grouped into a third class which we would suggest calling behavioral seasonality. These types of seasonality are discussed below.

2.2.1 Natural Seasonality

"Natural seasonality encompasses various factors that are directly related to the climate and its regular variations throughout the year (temperature, rainfall, snowfall, etc.)" (ALMEIDA; KASTENHOLZ, 2009, p. 6).

It can be said that the majority of tourist regions that suffer from seasonality are those that have a flow of visitors that varies according to thermal and temporal variations. However, it should be made clear that the climate is not the only motivating aspect of a trip, but it is a relevant factor that should always be taken into account.

The impact of climatic variations on the seasonal imbalance of tourist demand seems to be closely linked not only to the type of demand itself, but also to the existing structure of supply. While it is likely that a visit to any destination and its attractions will be all the more enjoyable the better the weather conditions, it seems equally true that some tourist activities do not depend exclusively on the weather for their realization, as is the case with cultural and urban leisure tourism. On the other hand, if the tourist offer in the destination area is structured around products or activities that are more dependent on certain climatic factors, such as snowfall in the case of winter sports, or pleasant temperatures and non-rainy weather for sun and beach tourism (and, in general, activities that take place outdoors), this will mean that, due to the absence of these factors at certain times, there will be no

27 KOENIG-LEWIS, N; BISCHOFF, E. **Seasonality Research**: The State of the Art. International Journal of Tourism Research 7, p. 201-219, 2005.

permanent availability of the offer, thus justifying the seasonality of the destination.

Although climatic variations throughout the year are important in explaining fluctuations in demand in certain destination areas (especially those where supply is more dependent on these conditions), they cannot override other explanatory factors of equal or even greater importance, such as institutional factors.

1.1.2 Institutional Seasonality

According to Butler[28] (1994 *apud* ALMEIDA; KASTENHOLZ, 2009), institutional seasonality is the terminology generally adopted:

> Fluctuations in tourism demand are explained by institutional factors, i.e. those that refer to human choices and decisions of a social and professional nature and which are usually rooted in customs, traditions and even legislation (ALMEIDA; KASTENHOLZ, 2009, p. 7).

These factors are at the root of opportunities and limitations to leisure and, in particular, to vacationing. It is suggested that the diverse set of factors underlying this category can be classified into two large groups according to their content: socio-cultural factors and socio-economic factors, as shown in Table 2.

Socio-cultural factors	**Socio-economic factors**
- Different calendars instituted:	• Professional leave
- Holidays	• School Holidays
- Religious calendar	• Economic constraints
- Calendar of pagan events	
- Cultural and sports calendar	

CHART 2 - GROUPING OF INSTITUTIONAL EXPLANATORY FACTORS SOURCE: ALMEIDA; KASTENHOLZ (2009, p. 7).

From the table above, it can be seen that various social, cultural and economic factors are just as important as those related to nature and climate. People travel when they have the opportunity and for some reason, and in this case the importance of vacations, school vacations, events and, consequently, the economic condition of the individual and family is evident.

1.1.2.1 Socio-cultural factors

The different calendars established in society are responsible for specific tourist flows at certain times of the year to different destination areas according to their specific vocations (ALMEIDA; KAsTENHoLZ, 2009). Thus, destinations dedicated to religious tourism (e.g.

28 BUTLER, R. **Seasonality in tourism**: Issues and problems. Tourism: The State of Art. A. V. Seaton. Chichester: Wiley, p. 332-339, 1994.

Aparecida do Norte - SP, or Jerusalem in Israel) record a significant flow of tourists on certain dates defined in the religious calendar, leading to the formation of inevitable peaks in demand in these places. Other specific dates and events, such as Carnival, lead tourists to visit destinations where the annual celebrations become major tourist attractions (e.g. Rio de Janeiro and Salvador).

Cultural or sporting events, which are regularly scheduled, also drive tourist flows to the destinations where they take place each year. To these types of dates, we must add the role of national and local holidays as drivers of tourist flows. According to Butler[29] (2001 *apud* ALMEIDA; KASTENHOLZ (2009):

> The growing importance of public holidays for tourism and for the phenomenon of seasonality results from the tendency to extend a significant part of them, promoting *short breaks* and motivating significant tourist flows in certain destination areas (ALMEIDA; KASTENHOLZ, 2009, p. 8).

Some holidays are also capable of mobilizing a whole flow of tourists to a given location, either to take part in events, or simply for rest and leisure away from home, and are responsible for a high turnover of people in the destination areas, such as national holidays, like Carnival, Easter, May 1st, Independence Day, among others.

1.1.2.2 Socio-economic factors

This group includes the factors that are perhaps most often mentioned in the literature with regard to institutional seasonality: professional and school vacations (CEC, 1993; COOPER, FLETCHER *et al.* 1993; CTP, 1998; BUTLER, 2001[30] , *apud* ALMEIDA; KASTENHOLZ, 2009, p. 8). These two factors directly and especially affect the working population in terms of leisure and vacation time. More specifically, limitations on the duration, timing and distribution of work and school vacations normally and formally affect tourism consumption and influence the concentration of demand in a destination area at a certain time of year.

Reflecting on how the working vacation period is used, it is clear that a significant percentage of the population still takes their vacations in a compressed way, i.e. without any distribution. However, the potential of interspersed short vacations to reduce seasonality is recognized.

According to Cockrell[31] (1998 *apud* ALMEIDA; KASTENHOLZ 2009, p. 8-9) the frequency of trips has been growing steadily, as tourists prefer to take two trips a year, rather than long concentrated annual vacations. This trend, accompanied by a general decrease in the average length of trips, began at the beginning of the 1980s, with a more pronounced growth

---

29 BUTLER, R. **Seasonality in Tourism**: Issues and Implications. Seasonality in Tourism. T. Baum and S. Lundtorp. Oxford: Pergamont, p. 5-21, 2001.
30 BUTLER, R. **Seasonality in Tourism**: Issues and Implications. Seasonality in Tourism. T. Baum and S. Lundtorp. Oxford: Pergamont, p. 5-21, 2001.
31 COCKRELL, N. **Outbond Markets/ Market Segment Studies**: The Short Break Market in Europe. Travel & Tourism Analyst - The Economic Intelligence Unit, 5, p.42-50, 1998.

in trips of less than 4 nights, related to the increasing use of public holidays (ALMEIDA; KASTENHOLZ, 2009). This growth is also pointed out by the UNWTO as one of the main tourism trends:

> Many tourists prefer shorter and more frequent vacations during the year. This offers the possibility for the development of new destinations and for existing destinations to offer tourists facilities and services for use in different seasons of the year (UNWTO, 1995, p. 21).

Although school vacations are rigidly defined in terms of their length and timing, they are generally spread over three or four periods a year, which, compared to the compacting of working vacations into a single period, could prove to be a positive influence. However, the longest school vacation period coincides with the summer (for reasons that have their origins in the need for extra labor for the harvest) (BUTLER[32] 1994, *apud* ALMEIDA; KASTENHOLZ, 2009, p. 9), which means that professional vacations are also scheduled at this time, with direct effects on the definition of seasonal demand patterns. It can be seen, then, that this situation is characteristic of both the Northern and Southern Hemispheres, which translates into the availability of tourist demand, in these two different origins, at opposite times of the year.

According to Almeida and Kastenholz (2009):

> Although economic factors are well known as general determinants of tourism demand, they have not yet been sufficiently studied in terms of their influence on the formation of seasonal demand patterns. Few authors even highlight the potential contribution of this factor as a cause of seasonality. However, it is important to argue the relevance of economic factors (income, prices, exchange rates, etc.) in determining consumption behavior, with consequences for seasonality (ALMEIDA; KASTENHOLZ, 2009, p. 9).

It can be seen from the above that natural and institutional factors are more important than economic factors when it comes to taking a vacation or traveling for a short period of time, but that economic factors become important when choosing one destination over another.

1.1.3 Behavioral Seasonality

This category groups together factors related to certain motivations and preferences of tourists that shape their consumption behavior and influence seasonality (ALMEIDA; KASTENHOLZ, 2009). Although almost all the factors reviewed can be read in this way, and can be seen as motivations for tourist consumption with psychological and behavioral roots, such as sun and beach tourism (natural seasonality) or the greater or lesser distribution of professional vacations (institutional seasonality), there are, however, other factors of this nature that can be considered independently.

32 BUTLER, R. **Seasonality in tourism**: Issues and problems. Tourism: The State of Art. A. V. Seaton. Chichester: Wiley, p. 332-339, 1994.

One of these factors, reported by Osborne[33] (1992 *apud* ALMEIDA; KASTENHOLZ 2009, p. 10) "consists of the habits and traditions that shape certain tourist consumption attitudes." The author suggests that this may be an important justification for seasonality in tourism, although there are not enough studies to prove it. "This factor translates into the maintenance of consumption behaviour which, due to inertia or even resistance to change, characterizes tourists who are no longer affected by the constraints of the past (e.g. professional vacations)." (ALMEIDA; KASTENHOLZ, 2009, p. 10). Although the timing of the vacation period may be one of the main aspects of this consumer behavior, it is also possible that the choice of how to divide up the vacations may also be influenced.

Another factor capable of influencing seasonal visiting and consumption behavior is the social pressure induced by fashion. According to Butler[34] (2001 *apud* ALMEIDA; KASTENHOLZ, 2009):

> In many societies, the privileged elites, who are usually opinion leaders, have nurtured the tradition of dividing the year into specific seasons in which it is considered socially necessary to take part in selected activities in certain places, becoming a fashion to follow. (ALMEIDA; KASTENHOLZ, 2009, p. 10).

A good example would be the winter tourism that takes place in the Serra Gaûcha and Serra Paulista (Campos do Jordao), which motivates strong tourist flows and, consequently, the concentration of demand during these periods in the destination areas, whose offer is geared towards winter leisure activities.

From a broader perspective, another factor stands out, related to the massification of tourism, which results from the preference or motivation shown by some tourist segments for this type of tourism. This is consumption at a time when the environment in the destination area is denser, busier and more socially active, as is the case in coastal areas during the summer months. The European Commission's report (CEC[35] 1993 *apud* ALMEIDA; KASTENHOLZ, 2009) states that:

> The buzz that exists in many destinations due to large numbers of visitors is an important attraction for many tourists who are difficult to find in the low season. However, some segments, notably those with special interests and senior tourists, may prefer to avoid the hustle and bustle associated with the high season (ALMEIDA; KASTENHOLZ, 2009, p. 11).

It can be said that this preference or specific motivation directly influences the concentration of demand, both geographically and in terms of time. However, as has been observed, not all

---

33 OSBORNE, D. R. **Seasonality and habit persistence in a life cycle model of consumption**. Modelling Seasonality. S. Hylleberg. Oxford: Oxford University Press, p. 193-208, 1992.
34 BUTLER, R. **Seasonality in Tourism**: Issues and Implications. Seasonality in Tourism. T. Baum and S. Lundtorp. Oxford: Pergamont, p. 5-21, 2001.
35 CEC. **Season Tourism**: Analysis of Experience, Suitable Products and Clientele. Brussels: Commission of the European Communities, DG XXIII - Tourism Unit, 1993.

tourists have the same preference in this regard. As mass tourism develops, there is a parallel trend, especially in the West, towards individualism in tourism consumption, as people look for different trips, itineraries and alternative activities. According to Montejano (1996), this type of tourist is also called minority or elite tourism, and is practiced by a segment of the market that travels to less frequented, peripheral or even isolated places at any time of the year. Therefore, it is likely that this restricted group, which has become increasingly important, will, in contrast to the mass tourist, have a lesser appetite for concentrated consumption over time, which is likely to contribute positively to lower seasonality.

It can be concluded from this analysis that the different types of tourist motivations are also an essential explanation for the concentration of vacations in a given period, and therefore for the explanation of seasonality.

2.3 SEASONALITY OF TOURISM IN COASTAL AREAS

The coastal environment is, in its land proportions, considered to be humanly safe and privileged to be pleasant, when it offers the combination of land, sea and sun, making the vocation of these areas for tourism undeniable, especially those that are urbanized. According to Almeida (2000, p. 72):

> The subliminal tourist attractions of coastal environments are sun and salt. In them reside the elemental desires for color, health, disposition and leisure. They add to the religious and even mythical aspects of washing the body and soul after the seventh wave (ALMEIDA, 2000, p. 72).

But there's a downside to all this: coastal tourism is seasonal in most places, i.e. it's exploited to the point of exhaustion in just a few months of the year.

Tourism transforms places, regions and landscapes by creating attractions for the flow of visitors. The significant whole of this creation is not in the places, regions or landscapes. For Luchiari (2000, p. 36), "it is structured in society which, by revaluing natural landscapes and regional cultures, becomes its voracious consumer, setting in motion the real and symbolic construction of tourist territories." Seasonality ends up valuing the aesthetics of natural landscapes and tourist attractions in general for just one period of the year, forgetting that these attractions remain in place so that they can be enjoyed at any time.

It is necessary to develop forms of leisure for all, in which everyone involved can benefit, at all times of the year, without harming other human beings or the environment. Argues Krippendorf (1989, p.25):

> Perhaps vacations and leisure time spent away from home can precisely be transformed into a field of learning and experiences for everyday life and take our minds off a situation that no longer satisfies us, but traveling to take advantage of an opportunity: the opportunity to be enriched, to exercise

> freedom, mutual understanding and solidarity, and to discover a little of all this in everyday life. (KRIPPENDORF, 1989, p. 25).

And taking advantage of an opportunity doesn't mean that it only has to be at a certain time of year, when everyone decides to take some time off for leisure and relaxation. There are many ways to attract visitors all year round, the question is how to give them the opportunity. What to do and how to plan to attract visitors all year round. Find out what visitors are looking for out of season.

Before trying to analyze seasonality and what can be done to reverse this situation, the ideal is to analyze the tourist vocation of the receiving area and the characterization and classification of tourist flows. For Beni (1998, p. 382), in relation to vocation, the study can be divided into: "climatic and hydrothermal, landscape, cultural, religious, sporting, folkloric, handicraft, scientific, congressional, business or business and gambling." In relation to characterization, Beni subdivides it into: "tourist traffic and tourist flow." (BENI, 1998, p. 383). With regard to the classification of tourist flows, Beni subdivides them according to table 3:

TYPE OF FLOW	DESCRIPTION
Incoming tourist flow	It is the group of foreign or national tourists who flock to a given geographical area to stay for a limited period of time. It is subdivided into: intercontinental international inbound tourist flow, continental international inbound tourist flow, national inbound tourist flow or internal tourist flow, regional inbound tourist flow.
Emissive tourist flow	It is the group of foreign or domestic tourists who travel from a given geographical area to one or more receiving areas. It is subdivided into: intercontinental international outbound tourist flow, continental international outbound tourist flow, national or domestic outbound tourist flow and regional outbound tourist flow.
Itinerant tourist flow	This is done by tourists traveling to one or more reception centers and not staying there for more than 12 hours, i.e. not staying overnight.
Tourist flow	It is the movement of tourists who go to one or more receiving centers and stay there for more than 24 hours, thus generating an overnight stay. It is subdivided into: holiday-weekly tourist flow, holiday-mid-seasonal tourist flow, sedentary-residential tourist flow, sedentary-residential-weekly tourist flow, sedentary-residential-weekly tourist flow.

TABLE 3 - CLASSIFICATION OF TOURIST FLOWS SOURCE: Adapted from BENI (1998).

It can be seen from this that coastal tourism can be characterized by various types of flow, but the most common are the receptive, itinerant and stay types. By analyzing the type of flow together with the type of demand, the planner will be able to see ways of studying seasonality and consequently propose actions that could reverse it or possibly minimize its negative impacts.

Each municipality or tourism company needs to establish a desired level of transactions with the desired demand. Sometimes, the level of demand can be low, equal or higher than the desired level of demand. In places like the coast of Paraná, there are variations in demand (insufficient, adequate, excessive, etc.) that require a greater effort from municipalities and organizations to balance these levels. The different demand situations are quite common in municipalities and companies located in coastal areas that offer tourism products and services. Below is a table which, according to Kotler (1994), shows the different demand situations.

Situation	Features	Action
Negative Demand	The market is in a negative demand situation if a large part of the market dislikes and even avoids the product.	Analyze why the market doesn't like the product and modify it, propose lower prices and promotion focused on changing the market's beliefs and attitudes.
Non-existent demand	When target consumers are uninterested or indifferent to the product.	Finding ways to connect the benefits of the product with people's natural needs and interests.
Latent Demand	Many consumers may share a strong need that cannot be satisfied by the products on the market.	Measuring the size of the potential market and developing effective goods and services that would meet demand.
Declining Demand	Natural process in the product life cycle.	Reverse the process, analyze the causes of the decline and determine whether it can be stimulated by changing the characteristics and communication of the product on the market.
Irregular demand	Seasonal demand, very characteristic of tourism, causing problems of optimum capacity or saturation.	Carry out synchronization marketing, finding ways to change the same pattern of demand through flexible pricing, promotion and other incentives.
Full Demand	Satisfactory level of turnover.	Maintaining the level of demand in the face of changing consumer preferences and competition, measuring the level of consumer satisfaction.

Excessive Demand	A higher level than you can or want to meet, known as *over-booking* in the hotel industry.	Promoting *demarketing*, finding alternatives to reduce temporary or permanent demand, by raising prices and reducing promotions and services; it does not aim to destroy demand, but only to reduce its level, shifting it if possible.
Unwanted Demand	Demand for unwanted products - harmful or damaging.	Organized effort to discourage consumption through high prices, reduced product availability and effective communication.

TABLE 4 - DIFFERENT DEMAND SITUATIONS SOURCE: Adapted from KOTLER (1994, p. 31).

According to the table above, it can be stipulated that in Guaratuba, there is mainly irregular demand, as each season has a different number and type of tourists, and excessive demand, as in high season, most places are unable to satisfactorily serve their consumers, causing problems in service and a lack of products for all those interested. These types of demand have different effects on the municipality, and the most felt effects are in the economic and social spheres.

According to Bahl (2004, p. 42), "the large flows of travelers prove the complexity of interactions arising from human circulation, associated with a different objective or theme."

2.4 IMPACTS OF SEASONALITY

Seasonal tourism is complex, broad and difficult to explain. Why does seasonality exist? Receiving centers are prepared for three months of work, but what about the rest of the year? These impacts are very damaging, because local residents work excessively to receive tourists during the high season (summer) and during the rest of the year they are practically sedentary.

Seasonality brings with it social, environmental, administrative and, above all, economic problems. A more in-depth study is needed to try to characterize this situation, which is so damaging to seasonal tourism areas such as Guaratuba, the subject of this study.

Studying this topic requires an in-depth study of socio-environmental geography. According to Mendonça and Kozel,

> The approach to environmental issues, in order to be carried out in depth and in the dimension of society-nature interaction, thus breaks with one of the classic postulates of modern science, namely that which establishes the choice of only one method for the elaboration of scientific knowledge. Such an approach requires both the application of methods already tried and tested in the field of various particular sciences, and the formulation of new ones. But this characteristic is not unique to the environmental approach; it reflects geography's own identity in many of its experiences (MENDONÇA; KOZEL, 2004, p. 136).

Analyzing the words of the aforementioned authors, it can be seen that tourism is also multidisciplinary and can be studied from various angles in order to reach a consensus on its importance in a broader context. Seasonal tourism needs to be researched and analyzed so that its main points can be characterized and then worked on scientifically.

2.5 INTERRELATIONSHIP BETWEEN THE CAUSES OF SEASONALITY AND MITIGATION STRATEGIES

With a view to minimizing the effects of seasonality, and particularly in consolidated tourist destinations that suffer from these effects every year (as is the case of the municipalities on the Paranà coast, and more specifically Guaratuba), the necessary tourism planning cannot fail to consider forms of development to benefit a more balanced and, consequently, more sustainable tourism growth. If we add to this a strong economic dependence on tourist activity in areas such as Guaratuba, then this need becomes imperative, particularly in order to boost the attractiveness of investments in the sector, with a view to attracting demand at different times of the year. With this in mind, investors in these destinations which are more strongly affected by seasonality will be able to assume a legitimate expectation of efficiency gains and real growth potential from a greater balance in the flow of demand at different times of the year.

In fact, if the reduction of seasonality in a destination is achieved by increasing demand in off-peak seasons, without jeopardizing the maintenance of numbers and growth rates in the high season, a double effect can be expected, i.e. greater efficiency of tourist activity in the high season and real growth in demand and revenue in the low season. From this more optimistic perspective, the phenomenon could be seen as a challenge or opportunity leading to more balanced/sustained tourism growth in the destination area.

Pursuing this challenge could involve implementing a strategy to diversify tourism demand (ALMEIDA; KASTENHOLZ, 2009), developed in two complementary and interrelated areas:

> The attraction of different origins and different tourist segments, whose profile shows strong potential for attracting tourists at times of lower tourist influx in the destination area. The work of identifying and then selecting these segments/origins should be based on the results of an in-depth study of the constraints on tourist visits/consumption at certain times (particularly in the low season) and, on the other hand, on the motivations for visits and consumption behavior at different times of the year in the destination area. (ALMEIDA; KASTENHOLZ, 2009, p. 3).

It is difficult to find a destination area where the seasonality of demand can be explained by just one of the factors listed. The justification for the existence of seasonality can be found in a study of tourists' motivations for consumption at different times of the year and the effects they have on the destination.

In order to better understand the interactions between the causes of seasonality, it is worth introducing an analysis of its causes. The concentration of demand in a destination can be explained predominantly by the influence of factors inherent to the destination itself, such as climate, infrastructure, etc., the origin of tourists, or even a correlation between the two. Butler[36] (2001 *apud* ALMEIDA; KASTENHOLZ, 2009) suggests that:

> It is the interaction between the forces that determine the natural and institutional elements of tourism seasonality, both in the sending and receiving areas, as well as the changes caused by the actions of the public and private sectors that create the pattern of tourism seasonality. (ALMEIDA; KASTENHOLZ, 2009, p. 12).

In other words, the seasonality that characterizes demand in a given destination can, in some cases, be essentially explained by factors related to the origin of the tourists. It can therefore be seen that a destination area that is not subject to climatic variations, or even one characterized by a stable climate, can register strong seasonal fluctuations in demand, since this seasonality is often not only linked to the climate, but also to the offer of leisure options and infrastructure in seasons other than the vacations or summer.

It is important to argue that climatic variations are not always the only factors affecting a destination; there are other factors that are not only found in the destination, but can also be found in the place where tourist flows originate. In this case, the greater the degree of seasonality, the more dependent the destination is on the motivations of the source markets.

The causes of seasonality can then be organized according to their primary impact on the origin or destination of tourists. For Lundtorp *et al*[37] (1999 *apud* ALMEIDA; KASTENHOLZ, 2009):

> They suggest classifying them as *push* factors - related to the origin or *pull factors - related to* the destination. This means that some of the factors tend to influence the tourist, "pushing" him towards the destination area, and others exert an attraction effect from the destination, which once again reinforces the notion of the influence of explanatory factors on tourist consumption behavior. (ALMEIDA; KASTENHOLZ, 2009, p. 12).

The author suggests that for tourists to visit a destination, they need to be attracted to it or "pushed" towards it and that there are factors that need to be taken into account both at the origin and at the destination. It is believed that these factors have been mentioned above, as they take into account natural seasonality (climatic factors), institutional seasonality (socio-economic and socio-cultural factors) and behavioral seasonality (motivational factors). In this

---

36 BUTLER, R. **Seasonality in Tourism**: Issues and Implications. Seasonality in Tourism. T. Baum and S. Lundtorp. Oxford: Pergamont, p. 5-21, 2001.
37 LUNDTORP, S. *et al.* **The Off-season is 'no-season'**: the case of the Danish island of Bornholm. Tourism Economics 5 (1), p. 49-68, 1999.

context, Butler and Mao[38] (1996 *apud* ALMEIDA; KASTENHOLZ, 2009, p. 12) refer to the influence of modifying the seasonal pattern on the part of the organizations responsible for managing the destination, since what is important are the strategic guidelines for mitigating seasonality according to the combinations between tourism planning, supply options and demand levels. According to Almeida and Kastenholz (2009, p. 13), the modification process is essentially materialized in three vectors: "the diversification of attractions at the destination, seasonal price differentiation and the diversification of source markets."

Lage and Milone (2000) suggest the importance of four strategies that can help modify seasonal patterns:

> The diversification of the product *mix*; the diversification of markets (in relation to tourist segments); the differentiation of prices and the influence of political decision-making, facilitating a greater distribution of the vacation period throughout the year, thus acting on the factors of supply and demand simultaneously (LAGE; MILONE, 2000, p. 124).

As can be seen, the literature review highlights the importance of the strategy of diversifying demand in order to mitigate seasonality, both in terms of origins and market segments. The targets of this strategy should, however, only be selected according to the results of appropriate studies in the destination area, which prove the real possibilities of its implementation. This could involve proving that there are different motivations for visiting and consuming in each season, as well as understanding the causes and consequences of visiting at certain times of the year, which result in the formation of temporal imbalances in the destination. The following study aims to consolidate these arguments and characterize the seasonality of tourism in the municipality of Guaratuba - PR, its causes and effects.

---

38 BUTLER, R.; MAO, B. **Seasonality in Tourism**: Problems and Measurement. Quality Management in Urban Tourism. P. E. Murphy. Chichester: Wiley, p. 9-23, 1996.

CHAPTER III - CHARACTERIZATION OF THE MUNICIPALITY OF GUARATUBA - PR

As a contribution to the empirical study that follows in Chapter IV, this chapter sought to characterize the municipality of Guaratuba for a better understanding of the data analyzed and obtained and the entire geographical and economic context in which it is inserted, i.e. the Paranà coast.

3.1 THE COAST OF PARANA

The appropriation of beachfronts is a recent phenomenon in the history of coastal settlements and stems from a specific interest in coastlines, not previously present as a cultural manifestation, which is the use of the beach-based form of occupation.

Invariably linked to nearby towns that concentrate the production of capital and where its frequenters, the vacationers, permanently reside, the use of this form of occupation in the form of seaside resorts has generated settlements that are characterized above all by the dominant presence of second homes, which are used temporarily, and by the longitudinal occupation of the shores, which configures narrow, elongated urban patches that tend to occupy the entire length of the beaches, as shown in figure 3.

FIGURE 3 - CONTINUED OCCUPATION OF THE PARANA COASTLINE

SOURCE: LANDSAT satellite image - INPE, adapted by SAMPAIO (2006, p.184)

In the state of Paranâ, the form of occupation based on seaside resorts began in the 1920s, but developed effectively from the 1950s onwards, through an intense process of appropriation that completely dominated the ocean shores to the south of Paranaguà Bay, which resulted in damage such as coastal erosion, the impairment of watercourses and bodies of water, the destruction of the landscape and the expulsion of fishing colonies, as well as urban problems such as road disorder and the poor environmental quality of settlements. (IPARDES, 2008).

This form of occupation in coastal areas is defined contemporaneously by the desire to bathe in the sea and to be by the sea, to "sunbathe", to walk, to relax, to play sports, to meet up, and the beaches are its place of realization, and in the summers, above all, its time of realization. It brings with it two determining characteristics, according to Sampaio:

> Firstly, the interest of the establishment near the beaches, which has resulted in the appropriation of their shores (which did not occur for other uses), and secondly, seasonality, which results in a concentrated presence at certain times - during holidays, in particular, but also on national holiday and weekends - and the emptiness most of the time of what it produces, in turn, the idleness of its built base - housing, commerce, services and technical and social infrastructures - in these absences, and, particularly in underdeveloped countries or regions, the overload and inability to meet peak attendance, with particularly serious consequences for the environment. (SAMPAIO, 2006, p.170)

In this context of seasonality, it can be said that a large part of the coastline of the

Paranà suffers from high demand during the summer season and idleness throughout the rest of the year, creating serious social, environmental, cultural, economic and administrative problems.

The coast of Paraná is located between parallels 25° and 26° south latitude and 48° and 49° west longitude (IPARDES, 2010). According to the same source, it has a total surface area of 6,061.2 km^2 , which represents just 3% of the surface area of Paranà, and a total population of 255,320 inhabitants in 2007 (IPARDES, 2010).

This coastline is characterized by its physical and biological diversity, which has influenced the way it has been occupied, conditioning its uses. In this way, an analysis of the interrelationships between the natural base, occupation and its uses can help to identify and characterize conflicts and trends, i.e. social conflicts, for example, and the tendency to characterize second homes, especially in beach towns.

According to Pierri *et al.* (2006), Paranà is a state of plateaus, which make up around 95% of its territory, its other two landscape units being the Serra do Mar and the coast, with only 5%. However, the author emphasizes that diversity is concentrated in these two units. There are also hills and mountains up to 1,600 m high, remnants of ancient plateaus, alluvial plains and fans, coastal plains, large coastal islands and large estuarine complexes (PIERRI *et al.*, 2006) (figure 4).

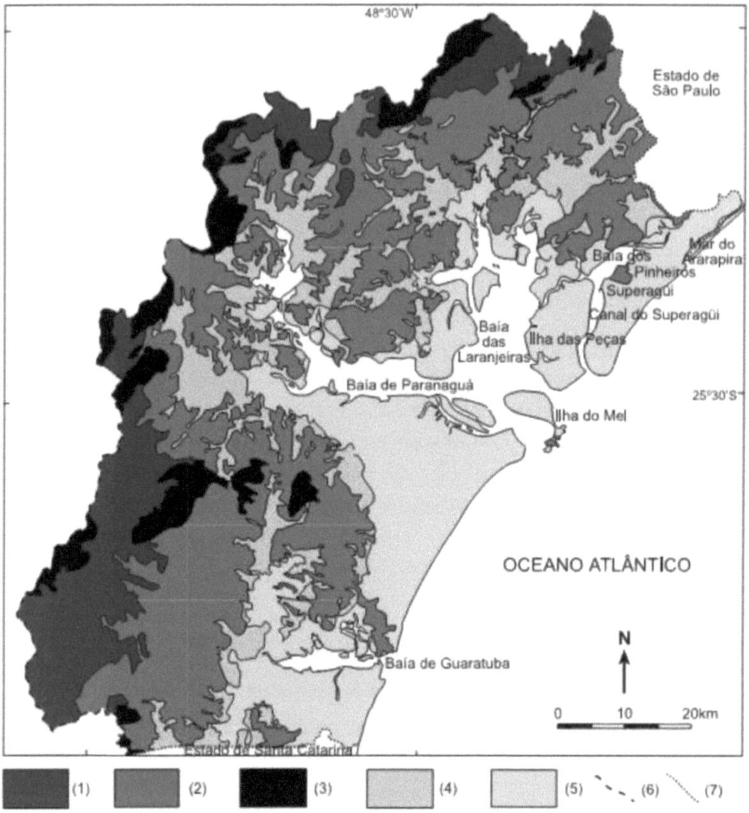

(1) PLANALTOS; (2) SERRAS ORIGINADAS POR DISSECAÇÃO DE BORDA DE PLANALTO; (3) SERRAS ORIGINADAS POR EROSÃO DIFERENCIAL; (4) TÁLUS, LEQUES ALUVIAIS E PLANÍCIES ALUVIAIS; (5) PLANÍCIE COSTEIRA; (6) DIVISOR DE ÁGUAS; (7) LIMITE INTERESTADUAL.

FIGURE 4 - MAIN GEOMORPHOLOGICAL UNITS OF THE COASTAL REGION OF Paraná.

SOURCE: PiERRi *et al.* (2006, p.140).

Angulo (2000) states that estuarine complexes give rise to a very jagged coastline and that while the ocean coast is around 100 km long, estuarine coastlines are approximately 1,300 km long. The same author also adds that the long coastline, combined with the geological, geomorphological and dynamic diversity of the coast, leads to the formation of numerous types of coastline. Considering the morphology and dynamics of waves, tides and associated currents, three main types of coast can be recognized (figure 5): estuarine, oceanic and inlet (ANGuLo, 2000). Pierri *et a.l* (2006) adds that:

> Ocean coastlines are dominated by wave action and are characterized by their sandy beaches and frontal dunes. Estuarine coasts are dominated by tides and

have tidal flats with a predominance of mangroves. Inland coasts are dominated by the action of waves and tidal currents and are characterized by their beaches and great mobility. In this type of coast there are frequent intense processes of erosion and sedimentation with large shifts in the coastline (PIERRI *et al.*, 2006, p.139).

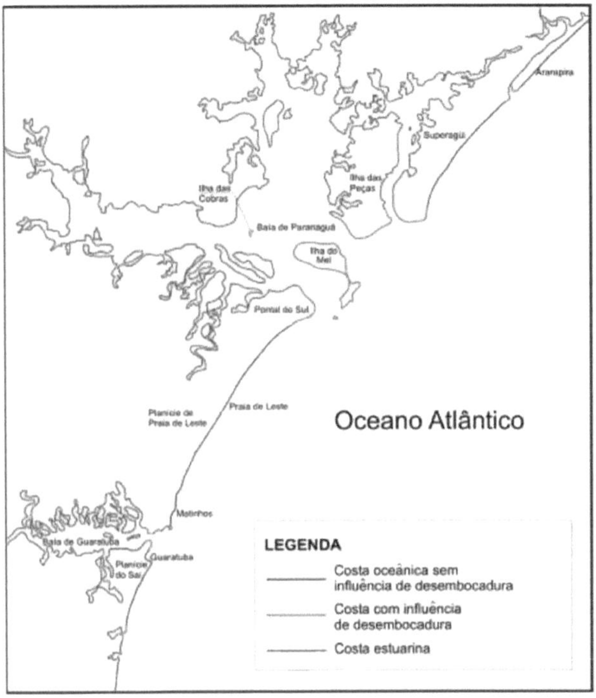

FIGURE 5 - TYPES OF COASTLINE. THE BEACHES CORRESPOND TO THE OCEAN COAST WITHOUT WAVE INFLUENCE AND THE COAST WITH WAVE INFLUENCE SOURCE: Adapted from Angulo and Araùjo[39] (1996) by SAMPAIO (2006, p.171).

In Paraná, the beaches are located along its Atlantic coastline and in some stretches of the mouths of its two bays. They total 126 km, distributed from northwest to southeast, according to IPARDES (2010): 31.0 Superagüi, 12.6 Ilha das Peças, 0.2 Ilha das Cobras (Paranagua Bay), 26.0 Ilha do Mel, 42.6 Continental edge between Paranaguà and Guaratuba Bays (Praia de Leste plain), 13.6 Continental edge to the south of Guaratuba Bay (Sai plain).

From an administrative point of view, the coast of Paraná is made up of seven municipalities: Guaraqueçaba, Antonina, Morretes, Paranaguà, Pontal do Sul, Matinhos and Guaratuba

39 ANGULO, R. J; ARAÙJO, A. D. de. **Classification of the coast of Paraná based on its dynamics as a subsidy to the occupation of the coastline**. Boletim Paranaense de Geociências, n. 44, p. 7-17, 1996.

(IPARDES, 2009). Until the middle of the 17th century, the entire area belonged to the state of São Paulo, and Paranaguà was dismembered in 1648, followed by the others, the last to be created being Pontal do Paranà, in 1997 (PARANAGUA, PREFEITURA MUNICIPAL, 2010). These municipalities are very close to Curitiba, the state capital: Antonina, the closest, is 63 km from Curitiba, and the furthest, Guaraqueçaba, 158 km (IPARDES, 2010). Antonina and Morretes, located at the bottom of Paranaguà Bay, do not reach the open sea, and Matinhos is the only one with a coastline exclusively on the open sea. Matinhos is the smallest, with 111.5 km^2 (2% of the total coastline); and the largest, Guaraqueçaba, with 2,159.3 km2 (35% of the total) (IPARDES, 2010).

It can be seen that the area and the permanent population are very unevenly distributed, with very different municipal densities. Paranaguà, with only 11% of the total area, has 54% of the inhabitants (139,796), while Guaraqueçaba, with 35% of the total area, has only 4% of the population (7,843 inhabitants) (IPARDES, 2010). The average population density is 38 inhabitants/km^2 , but varies between 4 inhabitants/km^2 in Guaraqueçaba, 24.69 inhabitants/km^2 in Guaratuba and 217 inhabitants/km^2 in Matinhos, followed by Paranaguà, with 191 inhabitants/km^2 (IPARDES, 2010).

As a result of this history, the shores of the beaches to the north of the Bay of Paranaguà - in Superagüi and Ilha das Peças -, above all due to the difficulty of access by land, remained on the fringes of the occupation process for beach use, which became established from the 1980s onwards, when they were included in Conservation Units, which cover almost the entire northern part of the coastal region (to the north of the Bay of Paranaguà) (PIERRI *et al*, 2006).

According to Paranaguà City Hall (2010), Ilha das Cobras has forbidden access and was once home to a prison (now deactivated). The site later became the official summer residence of the state governor, so its small beach is reserved for exclusive use.

Ilha do Mel has been frequented since the 1930s by upper-class families from Curitiba, but it was from the 1980s onwards that the pressure for its use increased, which is related to the island's appreciation as one of the most beautiful places on the coast of Paraná, but also to the improvement of its means of access and the availability of electricity (SAMPAIO, 2006). However, as far as land use is concerned, Paranaguà, Prefeitura Municipal (2010), the main area of the island is protected by Conservation Units, with only just under 5% of its total area earmarked for occupation.

According to Sampaio (2006), the course of occupation was the same along the different stretches of the coastline. As far as the type of settlement is concerned, they were always land subdivisions, in the form of allotments, called "balneàrios", with an almost absolute predominance of beachfront locations and, most of the time, without being continued, inland, by another development. The exceptions were notably in Matinhos, which grew inland as it became a city; between Matinhos and Caiobà, where the entire waterfront was occupied by the Cidade de Caiobà development, and near Guaratuba, where the first beach resorts merged with the pre-existing city (SAMAPAIO, 2006).

Towards the end of the 1970s, a set of problems arising from the use of the seafront began to emerge, and studies carried out by the state government identified their origin in the urbanization process that had taken place over the previous three decades (ANGULO, 2000). These problems, the diagnosis acknowledged, obviously stemmed from the extent to which seaside appropriation had reached, but above all from the disorder with which it was carried out, a situation that tended to worsen due to the pace of its growth (IPARDES[40] , 1980 *apud* SAMPAIO 2006, p. 176).

According to Angulo (2000), the results of land subdivision (and subsequent urbanization) included damage such as coastal erosion in Matinhos, Caiobà and Guaratuba (caused by the construction of roads and promenades on the beach), the risk of erosion in other places (caused by the occupation of stretches subject to the cycles of variation of the coastline), the impediment of the natural movements of the mouths of small rivers into the ocean (conditioned by the layout of the allotments), as well as the compromising of the landscape and the expulsion of fishing colonies, the latter due to the increase in real estate prices. The tall buildings in Caiobà and Guaratuba were damaging to the sunlight on the beaches, due to the shadows they cast for several hours a day; they also compromised the landscape by decontextualizing it, with consequences for the very use of the beaches; and the defensive works they imposed on marine dynamics, when located in areas affected by erosion, and which could themselves be reproducers of erosion processes. (ANGULO, 2000).

It can be seen, therefore, that the almost absolute absence of technical infrastructure and operational incapacity led to the pollution of rivers, bays and stretches of oceanfront by sewage and domestic waste, which would worsen with the proliferation of high-rise buildings.

According to Sampaio (2006), in the 1980s, the state government realized that the problems of occupation were taking on worrying dimensions on the coast of Paraná. At this point, the state created a document called "Padrões e normas técnicas para a ocupação e uso do solo no litoral paranaense" (Standards and technical norms for occupation and land use on the coast of Paraná) (IPARDES, 1980), which would be the first regional study produced after the 1966 one, but which, unlike that one, which aimed to plan for an occupation that was predominantly yet to take place, was now carried out in the contingency of the need to control an intense and disorderly process that already defined the entire seaside area.

Sampaio (2006) also argues that, among the causes of the situation diagnosed, the study

40 PARANAENSE INSTITUTE FOR ECONOMIC AND SOCIAL DEVELOPMENT - IPARDES. **Standards and technical norms for the occupation and use of land on the coast of Paraná**. Curitiba: State Planning Secretariat, 1980.

recognized the institutional fragility of the beach municipalities, whose conditions - precarious urban legislation and insufficient financial and human resources - placed them far below the set of forces that produced their spaces and, notably, their shores, where there was a clear predominance of interests linked to real estate production. As a consultative document, in addition to recommending a set of rules for land use and occupation planning in the Litoral Region (as a whole) to state government bodies, the study also proposed alternatives for direct interference by the state government in that territory (including in its urban areas), in order to impose criteria and controls that the municipalities were not practicing.

It emerged that all this work has attempted to curb disorderly growth without any planning, which has transformed the coastal region of Paraná into a lawless and highly speculative territory.

According to Sampaio (2006), one of the alternatives proposed - which suggested, on the basis of Federal Law 6.513/77, the classification of the Paraná coastline as a special area of tourist interest - prospered, and ended up being implemented, in the same year as the study was presented, by State Law 7.389/80, which thus made portions of the six coastal municipalities effective, and gave the state government powers over these areas, including the establishment of rules for their use, the function of giving prior approval to municipalities for land subdivision approvals, and also the right to supervise. Law 7.389/80 was followed by a process that culminated in 1984 with Decree 2.722, which is still in force and which regulated eleven categories of areas and sites of interest for protection purposes.

It is important to point out that the effectiveness of the state's intervention to control land occupation for seaside use on the coast of Paraná, more than twenty years after its establishment, is visible, especially with regard to the larger volumetric complexes, and, in a way, can be considered, given the time at which it took place (1984), as "a stroke of luck", which prevented the southern edge of the state from suffering a fate similar to that of other seaside resorts, where the absence of suitable construction parameters led to the well-known "walls" of buildings on the line facing the beach, which generated "heat islands" in the following blocks, due to the lack of ventilation.

According to Sampaio (2006), two years later, for reasons of national circumstances that completely escaped the motivations behind Decree 2.722/84 (the "Cruzado Plan" and the closure of financing for the real estate sector by the now defunct National Housing Bank - BNH), there was a strong redirection of investments by the real estate industry, which, seeing its housing markets reduced and dependent on long-term financing for their commercialization, they began to concentrate their actions (and their own capital) on the production of real estate aimed at the higher-income social strata, who were able to buy them without the support of financing, which made second homes in the most prized seaside areas an important segment to be exploited.

It is clear to see the importance of beach use on practically all of Paraná's coastline. According to Pierri *et al* (2006), beach use involves two major processes involving different populations and taking place at different rates, and to a certain extent, in different spaces. On the one hand, there is the influx of tourists, both those who build their second homes, driving urbanization closer to the sea and of better quality, and those who visit the coast without owning their own home. This is a population concentrated in the summer season and nearby holidays and, to a much lesser extent, at weekends during the rest of the year. On the other hand, there is a permanent influx of people who come from other regions to live on the coast, hoping to improve their income and quality of life, mostly low-income people, looking for new opportunities. They occupy less valued spaces, regularly or irregularly, and drive the most precarious urbanization, suffering day after day from the present lack of infrastructure and services.

An indicator of the first process is the flow of tourists during the season. Statistics from the Secretary of State for Tourism - SETU (2010) show that it reached an average of 1,518,826 people in the 2000 to 2006 seasons, with a minimum of 1,365,885 tourists in 2002, and a maximum of 1,643,892 tourists in 2005.

The set of data analyzed (SETU, 2010) shows that the tourist use of the coast mainly serves the residents of the state, particularly those who live in the capital, many of whom own their own homes there. This, coupled with the short distance and relatively good roads, allows them to frequent it assiduously.

According to Moura and Werneck (2000), beach use is characterized by the demand for beaches and their occupation. In Paranà, beach resorts were initially located on more sheltered beaches, where nearby hills or sandbanks reduced the energy of the waves. This was the case with the calm beaches of Caiobà and Matinhos, and the beaches of Guaratuba and Pontal do Sul.

Subsequently, the beaches with the highest wave energy were occupied until they formed the continuous occupation of the southern coast of Paranà. On the north coast, the process of beach occupation was slowed down by the lack of suitable land access and the difficulty of sea access. Pierri *et al.* (2006) argue that, in this sector, all the beaches are located on natural or artificial islands, as is the case of Superagüi; a peninsula that was isolated from the mainland by the excavation of the Varadouro canal in the 1960s. According to the same author, the potential for occupation was subsequently extremely reduced with the creation of the Superagüi National Park, which includes the beaches of Superagüi and Ilha das Peças, the Guaraqueçaba APA (Environmental Protection Area) and the Guaratuba APA. He also comments that fishing villages have been excluded from the park, notably Vila das Peças and Barra do Superagüi, which have come under increasing pressure from tourist occupation, as these villages provide access to the park's deserted beaches.

Finally, analyzing the words of Pierri *et al* (2006), the great expanse of preserved nature on the coast of Paraná, of more than economic environmental value, has made it possible, in the contemporary political context of concern for the protection of nature and biodiversity, to constitute most of its territory as a conservation area.

It can therefore be concluded that, at present, the state's coastline can be defined as a large conservation area (82% of its territory) (IPARDES, 2010) with a large export port. Therefore, both land uses do not serve the coastal population first and foremost: the former, theoretically, serves present and future generations; the latter, international trade, the economy of the country and the state and, in particular, the productive export sectors (SAMPAIO, 2006).

Secondarily, the coast allocates part of its land to bathing and fishing. The first is primarily for the enjoyment of citizens from Curitiba and other regions who choose this place to rest at certain times of the year; and the second, fishing, is the only one whose main beneficiaries are people who live on the coast all year round. (SAMPAIO, 2006)

Therefore, it can be considered that the main historically determined uses of the land on the coast of Paranà alienate its population from its main benefits, formulating once again the paradox of the contrast between natural wealth and social poverty, but also the contrast between so much social wealth of others, external to the region, and the poverty of the locals.

3.2 CONSIDERATIONS ON THE TOURISM MARKET ON THE COAST OF PARANA

The tourism market on the coast has played an important role in generating jobs and income and in distributing the region's population, making it a vector for the region's socio-spatial structuring.

According to IPARDES (2008), the Paranà Coastal Tourism Region has its economic hub in Paranaguà, whose economy is strongly linked to port activities, which are large-scale users of the services provided by the commercial establishments in the Tourism Production Chain. In the other municipalities, these services are intended to meet a local demand and another associated with leisure tourism. In Paranaguà, this explains the more intense use of the production chain's equipment in the low season months, giving a certain constancy to the rhythm of activities throughout the year, and its intensification in the summer months, with sun and beach tourism, which characterizes the seasonality of the chain's activities in the other municipalities in the region.

Thus, it can be said that two situations give particularity to tourism on the coast. According to IPARDES (2008), on the one hand there is seasonality, especially in the municipalities that depend on sun and beach tourism; on the other, the economic activities of the Port of Paranaguà, which, although they increase the size of general economic activity, distort the size of the tourism Gross Domestic Product (GDP) for the region as a whole.

Analysis of the study on the Paranà Coast Productive Chain - IPARDES (2008) showed that the coastal market is dominated by micro and small firms, run by their owners, whose main occupation is commercial activity. As a result, the vast majority of companies in the region are micro and small businesses with low capital reproduction capacity. Accommodation (hotels and inns) and food services are the main activities in the region's production chain. Although the situation is similar throughout the state, their dynamism in all the coastal municipalities is highly influenced by tourist activity.

IPARDES (2008) analyzed each of the activities in the state's Coastal Production Chain and was able to highlight some common characteristics in relation to the form of ownership, purpose of operations, scale of operations, links with the market and inter- and intra-chain relations:

- Form of ownership - the predominant type of company is individual, with family management; the managers tend to know most aspects of the company's administration, but not the workforce, and there is no general participation in the decision-making process; few establishments pass on part of the administrative processes to third parties.

- Purpose of operations - predominantly aimed at the local market; few target the regional market.

- Scale of operations - they operate on a small scale and, as a result, tend to have very limited participation in the market, as they are interdependent on the general system and the behavior of the global market.

- Link to the market - given their scale, on the other hand, they are independent in terms of business organization, as they are not part of a complex system, such as a small division of a large company. However, this independence means that the owners/managers have maximum authority and total control of the firm, although their freedom may be curbed by obligations to suppliers, financial institutions and the relationships required by the state and/or municipality; therefore, in this case, independence is a bad quality in terms of market relations, because the more independent the link, the less the company's capacity and interest in establishing cooperation and partnership relations with a view to expanding its market operations.

- Inter- and intra-chain production relationships - there are two flagships in the Coastal Tourism Chain: one, in the activity of Accommodation and Food Services, spatially located in the cities that offer sun and beach (Paranaguà (Ilha do Mel), Matinhos, Guaratuba and Pontal do Paranà); another, between accommodation and non-chain activities, as indicated by the administration of the Ports of Paranaguà and Antonina; and a third group that is sustained by the development of gastronomic and historical heritage tourism (located in

Morretes, Antonina and Paranaguà). On the other hand, two new fronts of interrelationship are beginning to emerge in the region, one linked to religious tourism and the other to activities focused on tourism in natural areas, or ecotourism. As has been pointed out, the asymmetry of the market conditions the establishment of networks and contributes to creating an environment of uncertainty, increasing the risks inherent in cooperative action between the various activities in the market. (IPARDES, 2008)

A further analysis of the research carried out by IPARDES (2008) reveals a lack of collective efficiency. This is a factor that contributes to restricting the offer of new attractions and tourism products that are formatted and competitive, in a system of cooperation between all the links in the chain. Although everyone - businesspeople, residents and tourists - agrees that the Paranà coast has a lot to offer in terms of tourism, there are complaints from the municipal authorities in relation to the state, from businesses in relation to both and the federal government, and from the community in relation to all of the above.

Within this study, IPARDES (2008) found a number of barriers preventing the strengthening of the Coastal Production Chain:

> a) poor management training and development; b) limitations in the training and development of human resources; c) limitations in information systems, lack of knowledge of the market and problems in reducing costs and marketing; d) lack of links with and ignorance of the use of new technologies, with little or no incorporation of technological innovation, including as a marketing tool for companies; e) difficult access to adequate lines of finance and credit under competitive conditions, due to ignorance of the existing credit mechanisms for micro and small companies. (IPARDES, 2008, p. 59)

These restrictions should not represent insurmountable limits to the development of the sector. The coast needs to value its space and not wait for the state government or just the city halls to do something. The entire community must work together to leverage the growth of the activity in the region, in order to minimize the negative effects of seasonality.

Through these studies, it can be seen that seasonality causes major social and economic problems and needs to be studied and, above all, analyzed in order to find solutions that minimize its impact on the various productive sectors of the coast, so as to bring a little more quality of life to the people who live there.

3.3 CHARACTERIZATION OF THE MUNICIPALITY OF GUARATUBA - PR

The need for studies related to seasonality becomes very relevant, as they can point to factors that can help tourist destinations prepare to receive tourists from all market niches, or from specific niches all year round, making life easier for the local population, who will feel their self-esteem grow as a result of the increased flow of tourists. But for seasonality to cease to exist, or at least be minimized, tourism needs to be planned in a conscious and orderly manner so as not to harm the environment in which it is inserted.

The study object for this research is the municipality of Guaratuba, which suffers the negative effects of seasonality every year.

3.3.1 History of Guaratuba

The history of the municipality that will be presented below contains events and dates that are briefly covered in order to understand the reasons and processes surrounding the beginning of the town, as well as the facts that have marked the history of Guaratuba over time. It's worth mentioning that historical information about the region is scarce and that the work "A Historia do Municipio de Guaratuba" by Joaquim da Silva Mafra (1952) was essential for discussing most of the issues related to the municipality's history.

The first reports about the coast of Paraná were made by Hans Staden, a German who came to Brazil bound for the Rio de la Plata. With the help of interpreters, he established the first contacts with the people of the region when his ship docked on the island of Superagüi (PDDU PONTAL DO PARANA, 2001).

According to Mafra (1952), D. Luis Antônio Botelho Mourâo, obeying the orders of the Marquis of Pombal - minister of D. Jose I, King of Portugal - Governor of the Province of São Paulo (to which Paranà was annexed), commissioned his nephew Afonso Botelho Sampaio e Souza, after years of new treaties, conflicts and invasions, to take charge of the territory of Paraná, with the following mission: to strengthen the coastline, preventing a Spanish attack. Thus, according to the author, the immediate steps taken were to fortify Ilha do Mel, building a fortress, and founding the town of Guaratuba. He goes on to say that these actions also sought to expand Portuguese territorial dominions, maintain control over the auriferous trade and facilitate the collection of taxes with the founding of the town.

According to Mafra (1952), in correspondence to the Marquis of Pombal, D.

Luiz Antônio writes, referring to Guaratuba:

> [...] and in favor of the same situation, I have also planned to found a settlement, fourteen leagues further down to the south, and in the cove called Guaratuba, because it is very abundant with fish, and with enough Barra, and has very small plots of land, for the said foundation, and there are also gold mines in them, so it is very necessary to have someone to defend them. (MAFRA, 1952, p. 5).

The story of the town's foundation begins, according to Mafra (1952), when Afonso Botelho, in compliance with D. Luiz's orders, summoned all the inhabitants of the region to gather in the small settlement that already existed in the Guaratuba cove. He brought together 200 couples, giving them the necessary tools and demarcating the land according to the needs and possibilities of each couple (MAFRA, 1952).

According to Mafra (1952), Dom Luis Antônio, approving the site chosen by his Adjutant of

Works, determined the location for the new town in a letter dated April 14, 1767:

> So I'd like you to go and see that stop, and fundamentally examine the place to found the town, because I think it's much better on the south side, because of its many advantages: firstly, because the place you've pointed out is capable of holding a town; thirdly, because the bottom of it (Guaratuba Bay) is capable of holding large ships. (MAFRA, 1952, p. 4).

Erected on April 30, 1771, the Pelourinho, according to Mafra (1952), was erected on the site of what would later become the Paço do Conselho e Cadeia, and later the Paço Sâo Luis. Mafra mentions it as a symbol of justice, as it was a column erected in the public square, where criminals were exposed to punishment. It consisted of a column made of massaranduba wood, with four sides measuring ten palms and from there upwards eight-pointed, with four iron rings, four quoins and a cleaver at the top (MAFRA, 1952).

When the town of Guaratuba was established in 1771, there were already many houses and bonfires were lit in the streets at night to provide lighting (MAFRA, 1952). On May 13, 1768, D. Luiz ordered the construction of a church that would serve as a spiritual center for the population. On April 29, 1771, he named the town Vila de Sao Luiz de Guaratuba da Marinha, and its solemnities included the celebration of Holy Mass. According to Mafra (1952), the first town council was elected on April 30, 1771, with the approval of the town and the general ombudsman.

According to Mafra (1952), SainfHilaire[41] , a French naturalist traveling along the Brazilian coast, described the city of Guaratuba in 1820:

> With about forty houses, fifteen or so form the semi-circle on the edge of the Angra; the others are behind them, around a vast grass-covered square, at the edge of which is the church. The oldest houses are nothing more than miserable shacks built of mud and in poor condition, but a short time before my trip, some beautiful stone ones had been built. (MAFRA, 1952, p.5)

With the proclamation of the Republic in 1889, according to Guaratuba, Prefeitura Municipal (2010), Guaratuba elected its first mayor, Senhor Manoel Antônio de Souza, who held the position until 1892, and the town continued in this legal status until October 20, 1938, when the municipality was abolished, becoming part of Paranagua as a simple district. Also according to data from the City Hall, on October 10, 1947, the municipality was restored its autonomy and Berilo da Cunha Padilha became the first mayor of this second administrative phase. The municipality of Guaratuba was part of the District of Paranagua until 1954, when it became the Judicial District of the District of Sao José dos Pinhais. The District of Guaratuba was created by Law 8280/86 and its installation took place on February 3, 1986,

41 Saint Hilaire, Augustin François César Prouvençal (1779 - 1853). He was a French botanist, naturalist and traveler. He traveled around Brazil for several years and wrote important books about Brazilian customs and landscapes in the 19th century.

through Ordinance No. 747/86 (GUARATUBA, MUNICIPAL PREFEITURE, 2010).

The name of the municipality of Guaratuba comes from the Tupi, which means "a lot of guarà" ("tuba", from the Indian, means "in excessive quantity"), according to Mafra (1952).

The guaras (figure 6) are scarlet-red seabirds with long, broad wings and very short tails, found in excessive abundance in the coastal region of Paraná (SEMA, 2001). According to the French naturalist SainfHlaire (*apud* MAFRA, 1952), these birds were also found in

in Santos and Santa Catarina, but spawning only took place on Ilha de Capim (formerly Guarà), in Guaratuba Bay. According to the same author, from August until November, they would gather in large groups, making crude nests in the branches of the mangroves, and their reproduction would be prodigious if the wind didn't knock most of them down, or if birds of prey didn't devour most of the eggs. The color of its feathers also attracted the greed of men and the hunting of this traditional and precious bird led to the almost total extinction of the guaràs (MAFRA, 1952).

FIGURA 6 - GUARÀ

SOURCE: GUARATUBA CITY HALL, 2009.

According to SEMA (2001), a law was passed in the 17th century prohibiting the molestation of these animals and their nests in the Guaratuba region, as the species was beginning to become extinct. Nevertheless, this measure was one of the first Brazilian governmental demands for nature conservation. Guarà populations gradually declined until the last record of this species in the state was made in 1979, when it was considered an endangered bird (SEMA, 2001).

According to Mafra (1952), there is a particularity that Saint Hilaire teaches[42] , in the commentary below, noted during his visit to the town:

42 Saint Hilaire, Augustin François César Prouvençal (1779 - 1853). He was a French botanist, naturalist and traveler. He traveled around Brazil for several years and wrote important books about Brazilian customs and landscapes in the 19th century.

> Boats between 80 and 100 meters long can enter it, but they don't dock in front of the city. In front of it there is a very narrow canal, and beyond it, one of those marshy islands covered in grasses called paratuva. It is on the other side of this island, which is very small and elongated, that the ships dock. I believe, moreover, that it will soon disappear, because every year, I'm told, the waters carry away part of it (MAFRA, 1952, p. 8).

The island referred to by the historian, according to Mafra (1952), is the part that was covered with paratura, where the pier rests (figures 7 and 8), and which bears no further resemblance to the island, blending into a single marshy plain that separates the canal from the plateau of Avenida Afonso Botelho (figure 9).

FIGURA 7 - VIEW OF THE PIER AND THE OLD PART OF GUARATUBA, TAKEN IN 1936 DURING THE HIGH TIDE.

SOURCE: Bigarella[43] , *et at.* (1970 *apud* PDDI GUARATUBA, 2002, p.14).

[43] BIGARELLA, J. J.; SILVA, J. X. da; DUARTE, G. M. **O desastre de Guaratuba: Um estudo de Geomorfologia Aplicada**. Curitiba: Impressora Paranaense S. A., 1970.

FIGURA 8 - VIEW OF THE ARRIVAL WALL AND TRAPICHE, BEFORE THE EROSION BEGAN.

SOURCE: Bigarella[44] , *et at.* (1970 *apud* PDDI GUARATUBA, 2002, p. 14).

FIGURA 9 - CONSTRUCTION WORK ON THE RETAINING WALL AND AVENUE CORONE AFONSO BOTELHO IN 1954.

SOURCE: Bigarella[45] , *et at.* (1970 *apud* PDDI GUARATUBA, 2002, p.14).

In this area, in addition to the old town hall building (figure 10), some houses (figure 11) and a pier were destroyed. The building that served as Guaratuba's town hall, which sank almost completely, caused major problems for the municipality, mainly due to the missing documents (PDDI, GUARATUBA, 2002). According to O Cruzeiro[46] (1968, *apud* PDDI GUARATUBA 2002), on the night of September 22 to 23, 1968, Guaratuba witnessed an

44

45

46

event that would forever be marked in the municipality's history. An area, as shown in Figure 12, about 130 meters long by 50 meters wide was literally "swallowed up" by the sea, with no deaths reported. Figure 12 shows a schematic of the area affected by the erosion.

FIGURE 1 o - OLD TOWN HALL, BEFORE THE SEA EROSION OF 1968. SOURCE: Parana em Paginas[47] (1978, *apud* PDDI GUARATUBA, 2002, p. 14).

FIGURE 11 - HOUSES DESTROYED IN THE DISASTER OF SEPTEMBER 23, 1968.

SOURCE: o Cruzeiro[48] (1968 *apud* PDDi GUARATUBA, 2002, p. 15).

47

48

FIGURE 12 - REPRESENTATIVE DIAGRAM OF THE SITE AFFECTED BY EROSION.

SOURCE: Bigarella[49] , *et at.* (1970, *apud* PDDI GUARATUBA, 2002, p.14)

According to Bigarella[50] (1970) *apud* PDDI Guaratuba (2002), the composition of the subsoil in the affected area is of particular importance in explaining the causes and processes involved in the Guaratuba disaster. The same author also argues that in front of the first buildings there were sandy and silty-clay sediments on which the pier and the stone retaining wall were built, about 3.5 meters high above the sediment surface. The author states that the waves (which have a wind-driven erosive effect), the tidal currents (which circulated at considerable speeds, becoming a powerful erosive agent, removing the natural protection), the water table (unbalanced in its hydrodynamic conditions) and the environmental variations introduced by human occupation were the set of factors that generated the process of sinking, collapsing and sliding in the area (figure 13).

[49] BIGARELLA, J. J.; SILVA, J. X. da; DUARTE, G. M. **O desastre de Guaratuba: Um estudo de Geomorfologia Aplicada**. Curitiba: Impressora Paranaense S. A., 1970.

[50] BIGARELLA, J. J.; SILVA, J. X. da; DUARTE, G. M. **O desastre de Guaratuba: Um estudo de Geomorfologia Aplicada**. Curitiba: Impressora Paranaense S. A., 1970.

FIGURE 13 - SITE OF THE 1968 EROSION

SOURCE: Paranà em Pâginas[51] (1978, *apud* PDDI GUARATUBA, 2002, p.15).

According to Guaratuba City Hall (2010), at that time a restoration project for the area was drawn up by Professor Arinos Xavier Tavares, consisting of a stone rockfill 180 meters long, whose axis is eleven and a half meters from the alignment of the original wall. Data from the City Hall says that the advantage of the rockfill would be that it would be more economical when compared to other possible solutions, guaranteeing that the process would not continue and allowing for almost total recovery of the lost area. But the project was not carried out and today the image of the site has not changed much since 1968, as shown in figure 14. Figure 15 shows the current image of the erosion site.

[51] PARANA EM PAGINAS MAGAZINE. Year XIV, no. 158. Curitiba, April 1978.

FIGURE 14 - CURRENT IMAGE OF THE SITE AFFECTED BY THE 1968 EROSION

SOURCE: The author (2010)

3.3.2 Main access roads

Guaratuba is interconnected with Paranà and other states in southern Brazil via the following state and federal roads (figure 15): PR-407; PR-412; BR-277. The construction and paving of access roads to Guaratuba began in the 1960s, creating new conditions for the flow and occupation of the coast (PDDI, Guaratuba, 2002). First, the Estrada do Café was completed and then, on April 6, 1968, the BR-277 was inaugurated, an access road to the coast of Paraná, facilitating access to the port of Paranaguà (PDDI GUARATUBA, 2002).

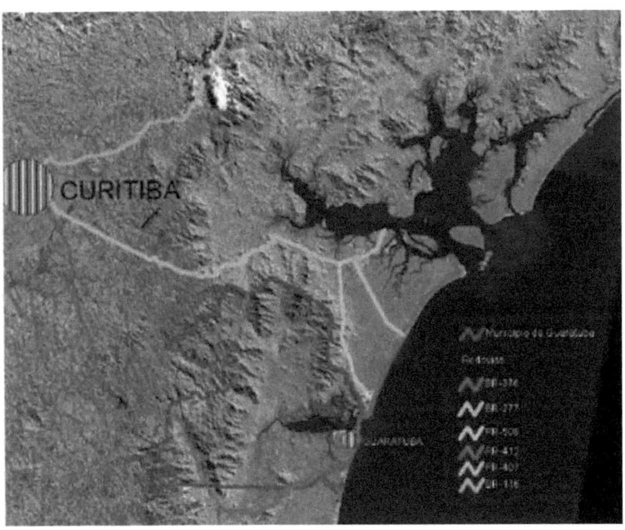

FIGURE 15 - REPRESENTATIVE SCHEME OF THE MAIN ACCESS ROADS TO GUARATUBA. SOURCE: PDDI GUARATUBA (2002).

According to Guaratuba City Hall (2009), the 1970s were an extremely important decade for the intensification of urban occupation and the flow of people into the municipality of Guaratuba. Still according to the City Hall, on January 1, 1970, the PR-407 was built, linking the BR-277 to Praia de Leste, and on May 10, 1971, the paving of the stretch of the BR-376 between Curitiba and Garuva (Santa Catarina) was inaugurated. The PR-412, which connects Guaratuba to Garuva and Guaratuba to Pontal do Sul (Paranâ), was made available in 1974 and is of significant importance to the municipality (PDDI, GUARATUBA, 2002). With its construction, easy access to the region was made possible, avoiding the *ferry* crossing, and, as an inductive axis, it contributed to the increase in urban occupation in the municipality. It connects to the BR-376 highway, which gives access to Santa Catarina and Rio Grande do Sul.

3.3.3 Geographical location

The municipality of Guaratuba is located in the southern region of the coast of Paraná (figure 16), on the map sheet Guaratuba MI - 2858-4, coordinates in the UTM projection system - 680.000, 740.000 and 7.120.000, 7.170.000, (PDDI GUARATUBA, 2002). According to data from Guaratuba City Hall (2009), the municipality is located at an altitude of 3m, has an area of 1,316.51 km^2 and is 115 km from Curitiba, the state capital, 65 km from Joinville, in Santa Catarina, and 54 km from the city of Paranaguà. It has 30,973 inhabitants (IBGE, 2007), 86% of whom live in urban areas and 14% in rural areas. Its climate is tropical super-humid, without a defined dry season, with an average temperature of 22°C in summer and 18°C in winter (IPARDES, 2010). Its latitude is 25° 52' 58" South and its longitude is 48° 34' 29" West (IPARDES, 2010). It is bordered to the north by the municipalities of Morretes and Paranaguà, to the west by the municipality of Sao José dos Pinhais, to the east by the municipality of Matinhos and the Atlantic Ocean and to the south by the municipalities of Itapoà and Garuva in the state of Santa Catarina.

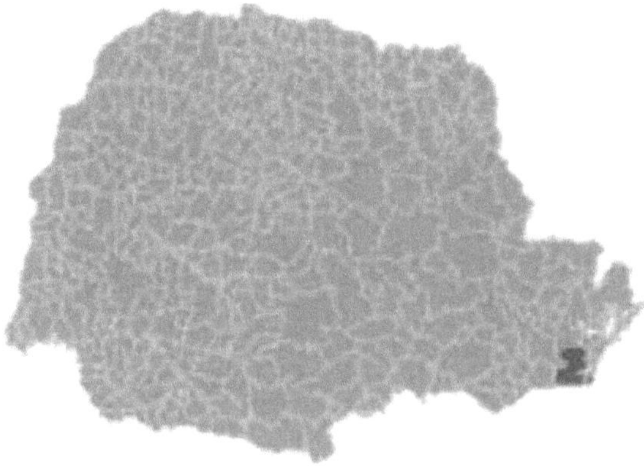

FIGURE 16 - LOCATION OF THE MUNICIPALITY IN THE STATE OF PARANA

SOURCE: IPARDES, 2009

FIGURE 17 - LOCATION MAP OF THE MUNICIPALITY

SOURCE: PDDI GUARATUBA (2002, p. 18).

In order to situate the municipality, it is necessary to include a satellite photo (figure 18), where the entire area of Guaratuba can be seen. The area in the center, in blue, is called Guaratuba Bay, a place much exploited by nautical tourism and fishing as a factor in the survival of the local population.

N

O 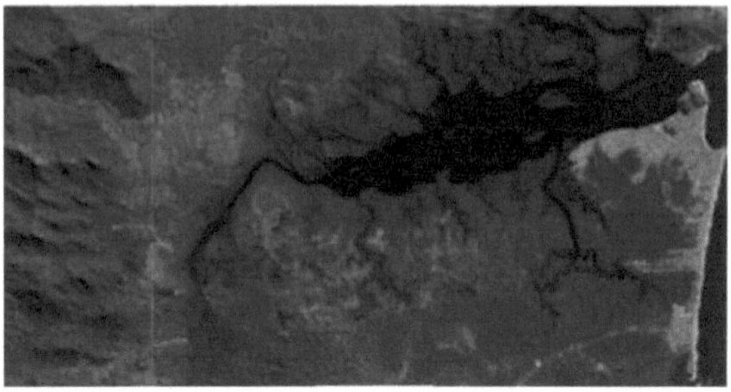 L

S

FIGURE 18 - SATELLITE PHOTO OF GUARATUBA BAY SOURCE: GuARATuBA, MUNICIPAL PREFEiTuRA (2007)

3.3.3.1 Geology

In Paranà, the Planicie Litorànea, or Coastal Plain, stretches from the foothills of the Serra to the ocean and is approximately 90km long and has a maximum width of around 55km in the Paranaguà region (ANGULO, 2000).

According to PDDI Guaratuba (2002), the plain is deeply cut by the estuarine complexes of the Paranaguà, Laranjeiras, Pinheiros and Guaratuba bays, resulting in numerous islands, such as the Peças, do Mel, Rasa, da Cotinga and Rasa da Cotinga islands, the plain is generally less than 20m above sea level. In several places, there are isolated hills and hills in the plain, with altitudes that can reach hundreds of meters, such as the Grande (479m) and Canudal (245m) hills and, in general, the height of the plain increases from the coast to the mainland, reaching the highest altitudes at the foot of the mountains (PDDI, GUARATUBA, 2002).

In general, the coastal plain is made up of continental and coastal sediments (GUARATUBA, SECRETARIA MUNICIPAL DE MEIO AMBIENTE, 2009). The former include sediments associated with slopes, such as fans, tàlus, colluvium and fluvial sediments. The continental deposits include two units that have been given formal stratigraphic names: the Alexandra Formation and the Iquererim Formation (GUARATUBA, MUNICIPAL ENVIRONMENT SECRETARIAT, 2009).

According to the Guaratuba PDDI:

> Coastal sediments belong to two main types of system: the coastal plain with coastal strands, and the estuarine system. Both are represented both by ancient environments formed during periods when the sea had relative levels higher than today, and by current environments such as beaches, tidal flats, tidal deltas and frontal dunes (PDDI GUARATUBA, 2002).

According to Ângulo (2000), in general the altitudes of the plateau are less than 20m and on isolated hills and hillsides, altitudes of up to 500m occur. The altitude of the plateau increases from the coast towards the continent, reaching the highest altitudes at the foot of the mountains. The plateau is made up of continental and coastal sediments (ANGULO 2000). The former include sediments associated with slopes, such as fans, talus, colluvium and fluvial sediments. Coastal sediments belong to two main types of system: the coastal plain with coastal strands and the estuary, both of which are represented by ancient and current environments (beaches, tidal flats, tidal deltas and frontal dunes). (ANGULO, 2000).

According to data taken from the Guaratuba PDDI (2002), coastal sediments were divided into two main groups, corresponding to different depositional systems. According to the same study, the first corresponds to the *strand plain,* which mainly includes *shore* face sediments, beaches and frontal dunes. The current environments of these systems include beaches, frontal dunes and intercordonal marshes (PDDI, GUARATUBA, 2002). In Paranà, the coastal plain with cordons extends from the Varadouro Channel in the north to the Sai-Guaçu River in the south, and in some sectors of the plain the cordons are poorly defined or have not yet been observed. They are often barely visible in the field, but are evident in aerial photographs and orbital images (PDDI GUARATUBA, 2002).

The second depositional system corresponds to the estuaries; in ancient sediments, tidal flats and shallow bottoms can be recognized, in present-day environments the tidal flats have been divided into six sub-units, and tidal deltas have also been recognized (PDDI, GUARATUBA, 2002). In terms of age, the coastal sediments were divided into Pleistocene and Holocene (IPARDES[52] ,1989 *apud* PDDI GUARATUBA, 2002).

3.3.3.2 Hydrography

The coastal region of Paranà is mainly made up of two river basins: the Paranaguà basin, which is approximately 3,882 km^2 long, and the Guaratuba basin, which covers an area of around 1,393 km2 (PDDI, GUARATUBA, 2002). These basins can be divided into different sub-basins. Completing the basins of the coastal region are the Sai-Guaçu river, to the south, on the border with the state of Santa Catarina, and the Ararapira Sea, to the north, on the border with the state of São Paulo, basins of small streams that drain directly into the sea, according to data from the Guaratuba Municipal Environment Department, 2009.

The Guaratuba hydrographic basin has rivers that originate in the Castelhanos, Prata, Igreja and Araraquara mountains (Cubatao river with its tributaries Cubataozinho, Arraial, Sao Joao, Guaratubinha and Castelhanos) and in the plain (rivers of lesser magnitude such as Pontal, Taquaraçu, Meio, Vitória and Claro) (PDDI, GUARATUBA, 2002). The Serra da Igreja, Serra da Canavieiras and Serra da Prata form a watershed draining into the bays of Paranaguà and Guaratuba, giving rise to most of the tributaries of the Cubataizinho river

52 IPARDES. Zoning of the Paraná Coast. Curitiba 1989b, 175p.

(Canavieiras, Furta Maré, Rasgado, Henrique, Guarajuran, Alegre das Onças) and coming from the south and flowing into Guaratuba bay, the Descoberto and Boguaçu rivers are also important (figure 19) (PDDI, GUARATUBA, 2002).

Between the Papanduras and Araraquara mountains rises the Sao Joao River, which runs through the Pedra Branca do Araraquara region and is bordered by the BR-376 highway in a north-south direction, passing through Garuva, in Santa Catarina, skirting the Morro da Onça region, and turning 180° returns to Paranà, already as a plain river, following a south-north direction until it flows into Guaratuba Bay (PDDI, GUARATUBA, 2002).

The same study also states that the interference of the territories of other municipalities in the Guaratuba hydrographic basin requires integrated actions for its management, especially in the case of the Cubatao River with its tributaries Sao Joao II, Arraial, Sao Joao River and, especially, the Sai-Guaçù River, since the latter, although it has a smaller territorial extension than the others, will serve as a source for the expansion of Guaratuba's water supply system (PDDI, GUARATUBA, 2002).

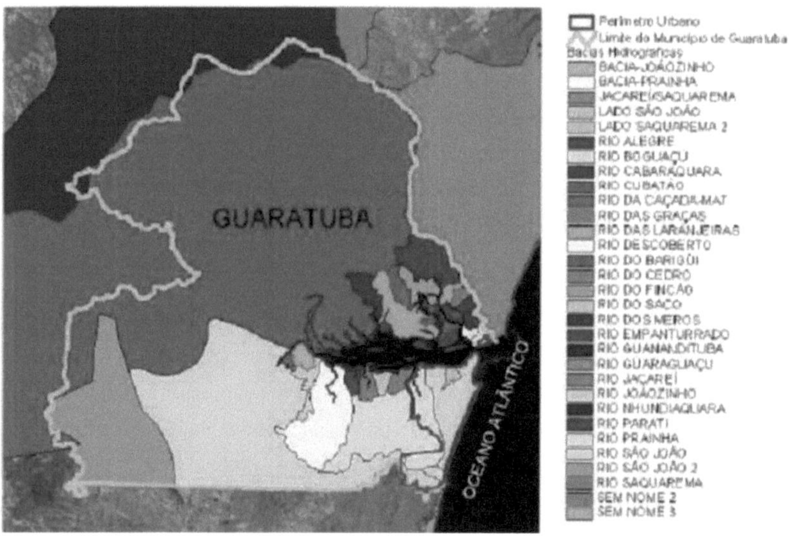

FIGURE 19 - WATERSHEDS IN THE MUNICIPALITY OF GUARATUBA. SOURCE: PDDI GUARATUBA (2002, p. 20)

3.3.4 Conservation Units

Certain natural landscapes have their value recognized by the degree of enchantment or uniqueness present. Lucas (1991) discusses the fact that the spirit of a place or genius 'loci' is the quality or qualities of the landscape that make it unique and special.

Paraná's coastline, whose natural heritage is remarkable and unique, had the Serra do Mar listed as the first institutional act to recognize its unique and special environmental value. On

July 25, 1986, the Serra do Mar was listed as a landmark in Paraná's preservationist culture (PARANAGUÂ, PREFEITURA MUNICIPAL, 2010).

With a total area of 376,000 hectares, the protected area begins at the junction of the Garuva - Cubatao highway, on the border between the states of Paraná and Santa Catarina, and includes areas in the municipalities of Guaratuba, Matinhos, Paranagua, Morretes, Antonina and Guaraqueçaba (PDDI, GUARATUBA, 2002). But as Rodolfo Angulo, who was responsible for the technical team that carried out the work to delimit the listed area, explained at the time: "The Serra do Mar has no precise boundaries, it is a regional geographical denomination that does not necessarily have a defined physical boundary."[53]

The Atlantic Forest Biome was internationally recognized in 1993. The United Nations Educational, Scientific and Cultural Organization - UNESCO, by decision of the Bureau of the International Coordinating Council of the Man and the Biosphere Program, created the Atlantic Forest Biosphere Reserve[54] , encompassing areas of the coast of Paraná, São Paulo and Rio de Janeiro. The area is part of the International Network of Biosphere Reserves, due to the well-deserved recognition of its biodiversity, historical and cultural identities and the ecological balance of the Serra do Mar and its landscape (SEMA, 2001).

On July 18, 2000, Federal Law 9985, regulating Article 225 of the Federal Constitution, established the National System of Nature Conservation Units. It establishes criteria and rules for the creation, implementation and management of conservation units. The law defines Conservation Units as:

> Territorial space and its environmental resources, including jurisdictional waters, with relevant natural characteristics, legally instituted by the Public Power, with conservation objectives and defined limits, under a special administration regime, to which adequate protection guarantees apply. (SEMA, 2001).

In view of the high quality of its landscape, the uniqueness of its bay, the beauty of its rivers, its biological diversity and the exceptional nature of the restinga and mangrove forest complexes that require protection, Guaratuba has three Conservation Units, according to the Guaratuba PDDI (2002): Environmental Protection Area, Boguaçu State Park, Saint Hilaire/Lange National Park (figure 20).

---

53 Extracted from the book 11Cadernos de Tombamento da Serra do Mar", Secretarla de Estado da Cultura. Curitiba, 1987, p. 84.

54 The Biosphere Reserve is an internationally adopted model for integrated, participatory and sustainable management of natural resources, with the basic objectives of preserving biological diversity, developing research activities, environmental monitoring, environmental education, sustainable development and improving people's quality of life. - Federal Law 9985/00.

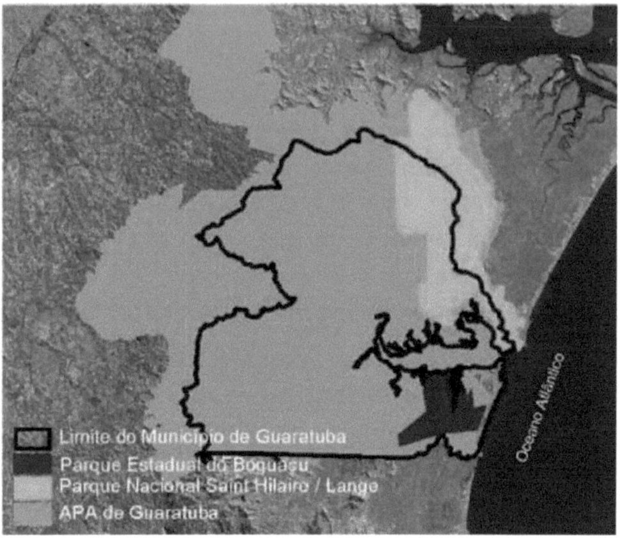

FIGURE 20 - LOCALIZATION OF CONSERVATION UNITS IN GUARATUBA SOURCE: PDDI GUARATUBA (2002, p.81)

At municipal level, in 1996, the site known as Lagoa do Parado and its tributaries was declared to be of public utility for the purposes of setting up the Lagoa do Parado Municipal Park (GUARATUBA, PREFEITURA MUNICIPAL, 2010).

3.3.4.1 Environmental Protection Area - APA Guaratuba

The Guaratuba APA, declared an environmental protection area on March 27, 1992 by State Decree 1234, aims to protect the biological, scenic and cultural aspects of an approximate area of 199,596.50 hectares of Atlantic Forest and associated ecosystems (figure 21). Its perimeter begins at the Serra do Mar Conservation Area and covers part of the municipalities of Guaratuba, Matinhos, Tijucas do Sul, Sâo José dos Pinhais and Morretes (GUARATUBA, MUNICIPAL PREFEITURA, 2010).

According to the Guaratuba PDDI (2002), the aim of the Guaratuba APA is to make the rational use of the region's environmental resources and the orderly occupation of the land compatible, to protect the water network, the remnants of the Guaratuba Forest and the surrounding area.

Atlantic and mangrove forests, archaeological sites and wildlife diversity. The aim of creating the Guaratuba APA was also to regulate tourist use and guarantee the quality of life of the caiçara communities and the local population. According to the same source, in the municipality of Guaratuba the APA covers approximately 98% of its territory, and in the urban area it affects all or part of 24 allotments. The allotments affected by the Guaratuba APA are: Jardim Village, Jardim Naçôes Bairro Fechado, Jardim dos Estados, Planta Piçarras, Jiçara,

Vila Rica, Castel Novo, Eliana, Jardim Rosana, Jardim Guaratuba, Nereidas de Guaratuba, Nereidas 2, Nereidas 3, Nereidas 4, Nereidas Novo, Cidade Balneària Brejatuba, Chàcaras Santo Amaro, Jardim Santo Amaro, Barra do Sai, Palmeiras, Jardim Nações I, Jardim Nações, Jardim As Nações, Parque Balneàrio Coroados.

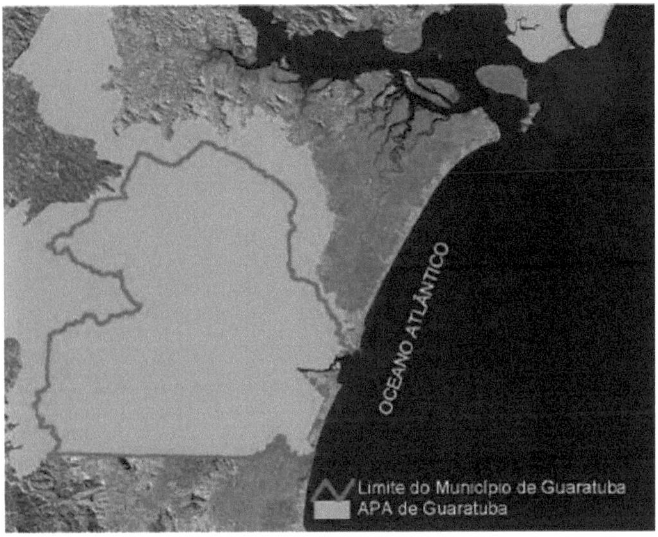

FIGURE 21 - LOCATION OF THE GUARATUBA APA

SOURCE: PDDI GUARATUBA, (2002)

According to the IAP (Environmental Institute of Paranà, 2010), the decree creating the APA establishes the need for zoning of use and occupation for the area, prohibits the implementation of potentially polluting industrial activities capable of affecting water sources and the estuarine complex of the bay. Also according to the same source, it is forbidden to carry out works or activities that alter ecological conditions, erode the land and/or silt up the water network, and prohibits the use of biocides when they are not in line with current techniques and standards. The IAP also requires an Environmental Impact Study and Environmental Impact Report (EIA/RIMA) to be drawn up for the opening of communication routes, canals and urbanization projects, with a view to controlling the polluting potential of activities for human use, and does not allow the construction of buildings that do not simultaneously include primary sewage treatment and the collection of water for supply, when there is no sewage collection and treatment network or water supply network. The IAP also prohibits the removal of sand and rocky material, nor does it allow construction of any kind, except for wharves, on marine and accrued land (IAP, 2010).

The APA of Guaratuba does not have a Management Plan[55] and the legal competence for its

55 According to Federal Law 9985/00, a Management Plan is a technical document which, based on

administration lies with the Directorate of Biodiversity and Protected Areas - DIBAP, of the Environmental Institute of Paranà. (PDDI, GUARATUBA, 2002)

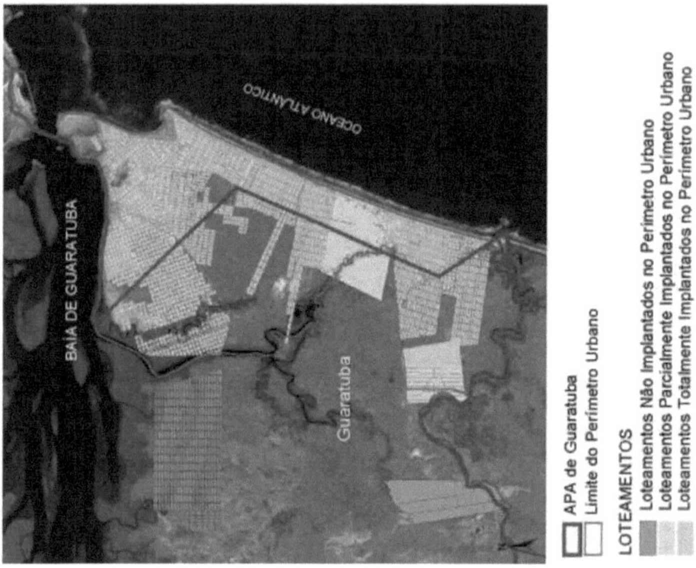

FIGURE 22 - LOCATION OF ALLOTMENTS IN RELATION TO THE GUARATUBA APA SOURCE: PDDI GUARATUBA (2002)

3.3.4.2 Boguaçu State Park

According to data provided by Guaratuba City Hall (2010), the park was created by State Decree 4056 of February 26, 1998 and covers an area of 6052 ha, with the aim of ensuring the preservation of the mangrove and restinga ecosystems around the Boguaçu and Boguaçu Mirim rivers. It is located on the edge of the current urbanized areas, including parts of 9 regularly approved allotments. The allotments affected by the Boguaçu State Park are Jardim Village, Jardim dos Estados, Jardim Rosana, Nereidas de Guaratuba, Nereidas 2, Nereidas 3, Nereidas Novo, Nereidas 4 and Cidade Balnearia Brejatuba (figure 23).

---

the general objectives of a conservation unit, establishes its zoning and the rules that should govern the use of the area and the management of natural resources, including the implementation of the physical structures necessary for the management of the unit.

FIGURE 23 - LOCATION OF BOGUAÇU STATE PARK SOURCE: PDDI GUARATUBA (2002)

In addition to the important function of preserving the mangrove and restinga ecosystems, this park could become an urban leisure and cultural facility, aimed at raising community awareness of the importance and notability of the coastal landscape.

According to the Guaratuba PDDI (2002), the State Park does not have a Management Plan and the legal competence for its administration lies with the Directorate of Biodiversity and Protected Areas - DIBAP, of the Paranâ Environmental Institute.

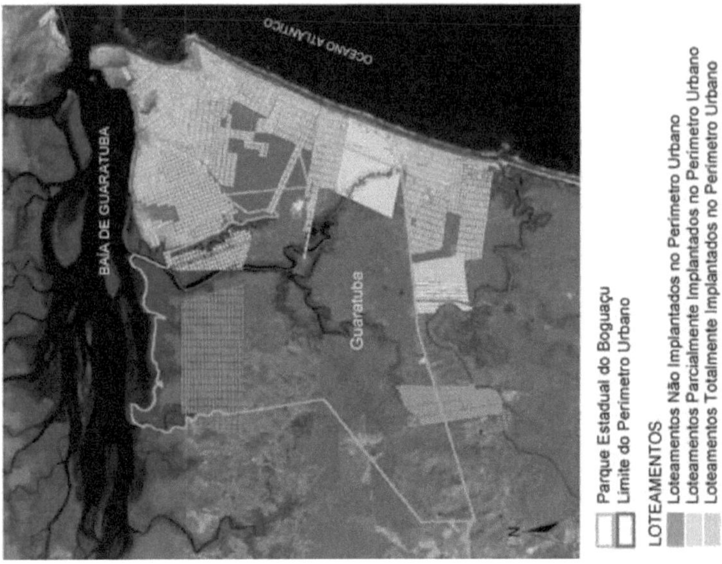

FIGURE 24 - LOCATION OF THE ALLOTMENTS IN RELATION TO THE STATE PARK OF BOGUAÇU

SOURCE: PDDI GUARATUBA (2002)

3.3.4.3 Saint Hilaire/Lange National Park

According to data provided by the Chico Mendes Institute for Biodiversity Conservation (ICMBio) (2010), the park was created by Federal Law No. 10.227, of May 23, 2001, and aims to protect and conserve the ecosystems of the Atlantic Forest and the environmental balance of the seaside resorts and communities. It also aims to protect public water supplies and preserve the quality of life of coastal populations. It is located in the southern portion of the Serra do Mar, in the Serra da Prata massif, at altitudes ranging from 50 to 1460 meters, with magnificent scenic beauty. Its vegetation system is Dense Ombrophilous Forest (Submontane, Montane and Altomontane). Located between latitudes 25°32'19" and 25°32'49" South and 48°4T52" and 48°32'23" West, it covers parts of the municipalities of Guaratuba, Matinhos and Paranaguà, with its administrative headquarters in Matinhos (ICMBio, 2010).

The Saint Hilaire/Lange National Park, which was created a short time ago, does not have a Management Plan. Its administration used to be the responsibility of the Brazilian Institute for the Environment and Renewable Natural Resources (IBAMA) and is currently the responsibility of the Chico Mendes Institute for Biodiversity Conservation (ICMBio, 2010).

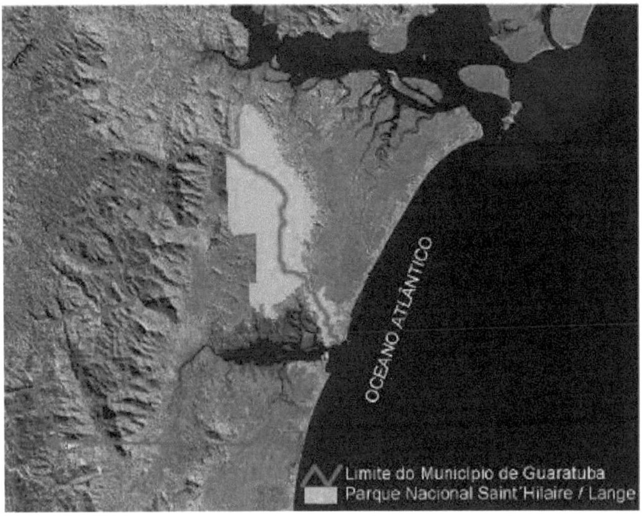

FIGURE 25 - LOCATION OF SAINT HILAIRE/LANGE NATIONAL PARK

SOURCE: PDDI GUARATUBA (2002)

3.3.4.4 Lagoa do Parado Municipal Park

According to data provided by Guaratuba City Hall (2010), on December 10, 1996, Municipal Decree no. 1626 declared the area covering Lagoa do Parado and its tributaries to be of public utility, for the purpose of establishing the Lagoa do Parado Municipal Park. Lagoa do Parado, considered a natural nursery for marine organisms, has a landscape of rare beauty. Located at the mean coordinates in the UTM projection system - 730,000 and 7,150,000 - the lagoon is part of the Atlantic Forest ecosystem (GUARATUBA, MUNICIPAL PREFEITURA (2010).

In order to effectively set up the Lagoa do Parado Municipal Park, it is necessary to establish specific regulations and define its perimeter. The declaration of public utility of the area covered by the lagoon is the milestone for the creation of an important municipal conservation unit. (PDDI, GUARATUBA, 2002).

FIGURE 26 - LOCATION OF LAGOA DO PARADO PARK

SOURCE: PDDI GUARATUBA (2002)

FIGURE 27 - AERIAL PHOTO OF LAGOA DO PARADO PARK SOURCE: PDDI GUARATUBA (2002)

3.3.5 Analysis of the basic infrastructure of the Municipality of Guaratuba

According to data provided by Guaratuba City Hall (2009), the following analysis was made: the central region of Guaratuba has the highest concentration of facilities and services, has a sewage collection system, water supply, paved and signposted streets.

Still according to the same source, the infrastructure in rural areas is made up of side roads with natural bedding, which make up a tangle of roads with little or no signposting, making it difficult to travel between the various towns in the municipality. It also shows that the rural schools, most of them elementary school from the first to the fourth grade, are located in the main communities: Descoberto, Alto da Serra, Caovi, Empanturrado, Limeira, Parati, Pedra Branca, Potreiro, Rasgadinho, Rio Bonito, Riozinho, Pirizal, Três Barras and Cubatao.

Guaratuba Bay and the long stretch of beach that runs from Caieiras to the southern edge of the municipality are the main tourist and leisure spots. These tourist attractions have led to urban growth along the beach and in the southern surroundings of the bay. This occupation is characterized by a markedly horizontal densification, with a low land use coefficient.

The water supply is collected from the Rio do Melo, located on Morro do Melo, and the municipal landfill is located in the basin of the Rio Boguaçu. The sewage treatment plant (ETE) is located in the south-central part of the urban area, between Avenida Paranà and Rua Afonso Pena (GUARATUBA, PREFEITURA MUNICIPAL, 2009).

In Guaratuba, the community of Cabaraquara, located to the north of the bay, has side roads that access the PR-412, which connects to the municipality of Matinhos, where the main services that the community's population uses are located (PDDI, GUARATUBA, 2002).

3.3.6 Socio-economic analysis of the municipality of Guaratuba

As a municipality that practically lives off tourism and the provision of services, the primary and secondary sectors are less important, but they do play a part in Guaratuba's economy. According to IPARDES (2010), the most significant agricultural activities are aquaculture, permanent crops (rice, bananas, sugar cane, beans, manioc, corn and tangerines) and livestock and other animal husbandry. As far as industrial production is concerned, the municipality receives *royalties* from the federal government, as a form of financial compensation for the production of oil and other minerals.

Guaratuba's industrial sector is present throughout the municipality through construction-related activities, which account for approximately 70% of the municipality's industrial establishments. There is also a slight diversification into the furniture industry, particularly joinery, metalwork and aluminum, and the manufacture of some food products (SCHEUER, 2010).

As for education, according to data from IPARDES (2010) based on a survey carried out in 2008, the municipality has 9664 students between nursery school and secondary school and approximately 350 students enrolled in the two higher education establishments (private), as shown in the table below:

ENROLLMENT IN BASIC EDUCATION AND PROFESSIONAL EDUCATION BY ADMINISTRATIVE DEPENDENCY-2008

DEPENDENCE ADMINISTRATIVE	CRECNE	PRE-SCHOOL	FUNDAMENTAL	MIDDLE	PROFESSIONAL

Federal		-	-	=	-
State	-	-	2.807	1.472	-
Municipal	245	777	2.852	-	-
Private	80	81	317	33	-
TOTAL	325	858	5.976	33	-

SOURCE: SEED

NOTES: 1 - Data refers to regular school enrollment with Inclusive 2 - Elementary School: includes enrollment in 8- and 9-year-old schools.

CHART 5 - STUDENTS ENROLLED IN GUARATUBA (KINDERGARTEN TO HIGH SCHOOL) SOURCE: IPARDES, 2010.

ENROLLMENTS AND COMPLETION BY ADMINISTRATIVE DEPENDENCY - HIGHER EDUCATION 2008

ADMINISTRATIVE DEPENDENCY	REGISTRATION	CONCLUINTES
Private	305	49
TOTAL	305	49

SOURCE: MEC-INEP

TABLE 6 - STUDENTS ENROLLED IN GUARATUBA (HIGHER EDUCATION)

SOURCE: IPARDES (2010)

As for supply, according to 2009 data, 21,823 inhabitants are served by Copel (electricity), 17,493 have treated water (Sanepar) and 7,964 have a sewage system connected to the system (Sanepar), which shows that less than half of the people who have treated water don't even use the sewage system, which may justify the pollution of rivers and seas (IPARDES, 2010).

F CONSUMPTION NUMFRODF ELECTRICITY CONSUMERS - 2009

CATEGORIES	CONSUMPTION (Mwh)	CONSUMERS
Residential	34.508	19.468
Secondary sector	2.928	375
Commercial sector	15.813	1.163
Rural	2.219	638
Other classes	11.851	178
Free consumption (Industry)(1)	24.184	1
TOTAL	91.503	21.823

SOURCE: COPEL$_h$ CQncessiQnarias - CPFL$_h$ COCEL$_h$ FORCEL$_h$ CFLO and

CELESC

(1) Refers to industry's self-production electricity consumption. It includes consumers served by another energy supplier and those who have part of their load served by COPEL Distribuçâo and the other part by another supplier.

TABLE 7 - ELECTRICITY CONSUMPTION AND NUMBER OF CONSUMERS

SOURCE: IPARDES (2010)

WATER SUPPLY$_h$ BYASANEPAR$_h$ BY CATEGORY - 2009

CATEGORIES	UN i T Y SERVED	LINKS
Residential	20.256	16.354
Commercial	1.062	908
Industrials	15	15
Public utility	119	116
Public authorities	101	IOO
TOTAL	21.553	17.493

SOURCE: SANEPAR

NOTE: Units (Ecoromies) Served is any property (house, apartment, store, building, etc.) or independent subdivision of the property with at least one water main, perfectly identifiable as an autonomous unit for the purposes of registration and tariff collection.

TABLE 8 - WATER SUPPLY BY SANEPAR

SOURCE: IPARDES (2010)

SEWERAGE SERVICE PROVIDED BY SANEPAR, BY CATEGORY - 2009

CATEGORIES	UN I D e D	LINKS
Residential	10.421	7.270
Commercial	721	597
Industrials	4	4
Public utility	-5	43
Public authorities	51	50
TOTAL	11.242	7.964

SOURCE: SANEPAR

A Served Premises is any property (house, apartment, store, building, etc.) or independent subdivision of the property, with at least one water main, which is perfectly identifiable as an independent Premises for the purposes of registration and tariff collection.

TABLE 9 - SEWAGE SERVICES PROVIDED BY SANEPAR

SOURCE: IPARDES (2010)

Guaratuba has the 5th highest HDI-M on the Paranà coast, corresponding to 0.764 (2009) and a population density (2009) of 24.69 inhabitants/km^2 . (IPARDES, 2010), as shown in the following tables:

Municipality	HDI-M	Ranging - Paranà
Matinhos	0,793	32°
Pontal do Paranà	0,788	41°
Paranaguà	0,782	58°
Antonina	0,770	91°
Guaratuba	0,764	108°
Morretes	0,755	139°
Guaraqueçaba	0,659	393°

TABLE 10 - *RANKING OF* THE IDH-M ON THE PARANA COAST.

SOURCE: Adapted from IPARDEs (2010)

DEMOGRAPHIC DENSITY - 2009

DEMOGRAPHIC DENSITY (inhab.⅛ın)ʲ

24,6 S

SOURCE: IPARDES

NOTE: Calculated based on the population estimated by IBGE

TABLE 11 - DEMOGRAPHIC DENSITY OF GUARATUBA SOURCE: IPARDEs (2010)

According to Scheuer (2010), commercial activities predominate in the food trade, accounting for almost 50% of the total number of commercial establishments registered in the municipality, and the construction sector stands out, represented by establishments selling building materials, given the continuous investment in construction and renovation of buildings. Local businesses include supermarkets, pharmacies, clothing stores, beachwear and gas stations (SCHEUER, 2010).

In terms of the services on offer, there is greater diversification, although the nature of the service provided is in line with the general characteristics of the municipality. In this sense, services focused on accommodation and food, repair, maintenance, cleaning and public entertainment predominate.

According to Scheuer (2010), the tertiary sector, which is the basis of the economy, is structured to cater for summer activities such as building materials stores, condominiums, bars and restaurants, bakeries, aviaries, auto parts, drinks distributors, butchers, pharmacies, video rental stores, gyms, etc.

The greater participation of women in the labor market is another important change in employment. The inclusion of women is mainly due to the need for families to supplement their income, which was previously restricted to the activities of the head of the household.

The community's involvement with tourism in Guaratuba began in the 1950s and 1960s, when the first vacationers from Curitiba and the north of Paranà began to visit the beaches on the state's coast, following the construction of highways such as BR-277 and BR-376 (PDDI, GUARATUBA, 2002). From this moment on, the interrelationship between tourists and the local community began, as second-home tourism began, mainly in the provision of services, such as gardening, domestic services, home-made products, fishing products and small services. (SCHEUER, 2010).

Currently, the municipality has a very heterogeneous population in terms of the origin of its inhabitants, so the community is less involved with tourists due to the veritable invasion of small and medium-sized traders who carry out their activities during the summer season, without giving the local community the opportunity to work and without taking advantage of their local knowledge (SCHEUER, 2010).

3.3.7 Tourism and leisure

Guaratuba has great tourist potential. Its landscapes and ecosystems of relevant biodiversity are considered by current legislation to be special for protection purposes (PDDI, GUARATUBA, 2010). The ocean, the strips of beach, the bay, its estuary complex, the mountains, the hills and the exuberant vegetation make up a scenery of great beauty and, at the same time, indicate a territory that is mainly used for tourism and the activities linked to it.

As Guaratuba has a large APA (environmental protection area) in the middle of the Atlantic Forest, it has preserved some areas of restinga on the edge of the beach which, as well as being home to typical plant species, serve as a nursery for some animals such as the burrowing owl (PDDI, GUARATUBA, 2002).

In Guaratuba, the main activity of mass tourism is leisure and recreation on the 15 km of beach, i.e. "sun and beach tourism", which stretches from Caieiras to Barra do Sai, including Prainha, a beach resort located to the north of Guaratuba Bay (SCHEUER, 2010). With stretches ranging from 6 to 280 meters wide, the beaches are the biggest attractions for tourism and generate economic activity during the summer season in the municipality (PDDI, GUARATUBA, 2002). Guaratuba also has a diversity of environments whose tourist potential is of great value. The estuary complex of the bay, and in particular the Boguaçu River and its tributaries, stand out for their importance. Notable are the Sai-Guaçu River, the Sâo Joâo River and the Parado Lagoon (PDDI GUARATUBA, 2002).

In terms of tourist attractions, Guaratuba is a city with many beaches, local events, religious sites, rivers, waterfalls, nautical, ecological and even extreme tourism. Below are some of the attractions to illustrate the town's potential.

3.3.7.1 Caieiras, Encantadas or Amores Beach

According to data from Guaratuba City Hall (2007), Caieiras beach is located between the tip of Johnscher and the rocks of Caieiras (Praia do Prosdócimo), with road access via the PR-412 and waterway access via the bay. It is 1 km long and is suitable for spearfishing. The same source suggests that the neighborhood originated from a fishing colony, of which there are still a majority. From the rocks, when the tide is low, you can see the prow of a steamer that ran aground while fleeing the Paraguayan War - Vapor Sao Paulo - commanded by the husband of composer and pianist Chiquinha Gonzaga. (GUARATUBA, MUNICIPAL PREFEITURA, 2009).

FIGURE 28 - CAIEIRAS BEACH

SOURCE: GUARATUBA, CITY HALL (2009)

3.3.7.2 Prosdocimo Beach

It starts at the rocks of Caieiras and runs as far as the Villa Real Hotel, the beginning of Avenida Atlântica. Its length forms a half-moon and continues towards Morro do Cristo (GUARATUBA, PREFEITURA MUNICIPAL, 2007).

FIGURE 29 - PROSDÓCIMO BEACH

SOURCE: The author (2010)

3.3.7.3 Central Beach

It starts at the Villa Real Hotel and extends to Morro do Cristo, forming the beaches of Cristo, Pescadores and Turcos. A half inlet beach, located at the end of Avenida Atlântica. Ideal for bathers, as the sea is calm. This beach is home to Morro do Cristo, which is very popular with tourists (GUARATUBA, PREFEITURA MUNICIPAL, 2007 and 2009). In 2009, Praia Central underwent a small revitalization process, but it is currently (2010) suffering from a hangover that has destroyed a large part of its retaining wall and it is without a strip of sand for bathers and hikers (figure 31).

FIGURA 30 - CENTRAL BEACH

SOURCE: The author (April, 2010)

FIGURA 31 - CENTRAL BEACH

SOURCE: The author (May 2010).

1.1.1.4 Brejatuba Beach

According to Guaratuba City Hall (2009), this is an open sea beach, with very rough waters and preserved natural vegetation, in a southerly direction after Morro do Cristo with 14 km of beach where each bathing resort is named: Estoril, Bonança, Eliana, Nereidas, Cidade

Balnearia, Santa Helena, Coroados and Barra do Sai, on the border with Santa Catarina. Brejatuba beach stands out for the surfing championships held there throughout the year, with average waves of eight feet; there is also fishing (GUARATUBA, PREFEITURA MUNICIPAL, 2009).

FIGURE 32 - BREJATUBA BEACH.

SOURCE: GUARATUBA, CITY HALL (2007).

1.1.1.5 Brejatuba Hill / Christ Hill

The inauguration of the statue of Christ on the Brejatuba/Cristo hill, according to Guaratuba City Hall (2009), took place on June 23, 1953. The image was cast by Joao Fedatto and his son Iswaldo, who, starting from a photograph, made a one-meter-high clay mold, followed by another made of plaster, and cement for the final work. The statue was made in ten rings so that it could be carried to the top of the hill more easily (PDDI, GUARATUBA, 2002). According to data from Guaratuba City Hall (2009), three of these rings collapsed during transportation on wooden hoists and were taken to Curitiba to be repaired. According to the same source, after this stage, the image was retouched and made with cement, having already been enlarged to its final size. The statue was a gift and a tribute to Professor Joao Càndido Ferreira's widow, Mrs. Josefa Amaral Ferreira (her participation in the execution of the work was one of the most important) (PDDI, GUARATUBA, 2002). Josefa's entire family took part in the inauguration ceremony, showing great enthusiasm for what was being offered to the city of Guaratuba. The event took place in the year of the centenary of the political emancipation of Paranà, and the governor at the time was Bento Munhoz da Rocha Netto (figure 33) (GUARATUBA, MUNICIPAL PREFEITURA, 2009).

FIGURE 33 - PHOTOGRAPH OF THE EVENT HELD ON JUNE 23, 1953, AT THE TOP OF CRISTO HILL. LADY JOSEFA DO AMARAL FERREIRA, BETWEEN HER grandchildren FRANCISCO CUNHA PEREIRA FILHO AND CECILIA FERREIRA SIMAS.

SOURCE: Parana em Paginas[56] (June, 1997 *apud* PDDI GUARATUBA, 2002).

The revitalization was completed on May 13, 2000. During this work, the image was moved from its original base (no longer facing Guaratuba Bay, but towards the city), gaining a 6.5-meter-high pedestal, with a memory room, oratory and belvedere (PDDI GUARATUBA,

56 Parana em Paginas magazine, June 1997.

2002). According to data from Guaratuba City Hall (2009), more than 900,000 people have visited this attraction since its inauguration and a vial containing holy water has been placed at the heart of the image.

A natural viewpoint, 38 meters high, it offers a spectacular view of the Atlantic Ocean and a privileged view of the entire region. Access to the hill is via a 199-step staircase (GUARATUBA, PREFEITURA MUNICIPAL, 2009). At the top of the hill, the image of Christ the Redeemer, 8.30m high, blesses the city. The image has one arm outstretched towards the city and the other over his body, with his hand on his heart (GUARATUBA, PREFEITURA MUNICIPAL, 2009).

FIGURE 34 - CRISTO HILL

SOURCE: The author (2010)

1.1.1.6 Our Lady of Lourdes Square

Located on the slopes of Morro do Pinto, in Rua Nossa Senhora de Lourdes, opposite the 29 de Abril State School. According to data provided by Guaratuba City Hall (2009), the site has a source of pure, crystal-clear water that supplied the community for several years until the treated water supply network was set up in 1974. Still according to the same source, it is estimated that more than 840,000 liters of this water flow from its spout every month, around 10 million a year and, as far as anyone knows, there has never been a shortage of water from this source. It is also called Itororó, which comes from the indigenous dialect: Itararé, meaning underground watercourse (GUARATUBA, PREFEITURA MUNICIPAL, 2009).

According to Guaratuba City Hall (2009), the image of Our Lady of Lourdes was donated by devotee Guilhermina Cordeiro and has blessed the square since July 30, 1935. According to the same source, many graces have been attributed to the saint, as evidenced by the various plaques that have been placed in the square.

by devotees in the chapel that protects the image, built by the couple Heitor and Dona Zina. People from all over Brazil and abroad come to Guaratuba to pay their respects and make promises to Our Lady of Lourdes for the graces they have received (GUARATUBA, PREFEITURA MUNICIPAL, 2009).

FIGURES 35 and 36 - LARGO NOSSA SENHORA DE LOURDES SOURCE: GUARATUBA, PREFEITURA MUNICIPAL (2008 and 2010).

1.1.1.7 Guaratuba Bay

According to data from Guaratuba City Hall (2009), Guaratuba Bay is an estuary embedded in the coastal plain of the Paranà coast and is the second largest in Paranà, measuring 48.72 km^2 in length. It is accessed by the *ferry* between Matinhos and Guaratuba. According to the same source, Guaratuba Bay is suitable for boat trips, fishing and water sports, is rich in fauna and flora and is an environmental protection area. The bay was the only access the community had to "Vila de Guaratuba" at the beginning of the municipality's colonization (MAFRA, 1952). The area upstream of the bay is intensively farmed with high-tech banana plantations (GUARATUBA, PREFEITURA MUNICIPAL, 2009).

FIGURE 37 - GUARATUBA BAY

SOURCE: GUARATUBA, CITY HALL (2008)

1.1.1.8 Parati jump

According to data from Guaratuba City Hall (2009), at the bottom of Guaratuba Bay, at the foot of the Serra do Mar, there is the Parati River, flanked by totally preserved Atlantic forest. The best way to get there is by sea, using small boats, and trips to the Salto are made all year round (especially during the summer) (GUARATUBA, PREFEITURA MUNICIPAL, 2009). The boat trip takes 25 minutes from the center across the bay to the Parati River, docking at a place called Porto (GUARATUBA, PREFEITURA MUNICIPAL, 2009). You then need to take a 40-minute walk along trails through native vegetation and typical coastal landscapes (GUARATUBA, PREFEITURA MUNICIPAL, 2009). The path passes several houses of local residents and a Sambaqui - deposits of shells, kitchen remains and skeletons accumulated by indigenous tribes that inhabited the coast, some of them more than 5,000 years ago (GUARATUBA, MUNICIPAL PREFEITURA, 2009).

FIGURE 38 - PARATI JUMP

SOURCE: GUARATUBA, CITY HALL (2010).

1.1.1.9 Largo do Carioca

According to data provided by Guaratuba City Hall (2009), the square is located at the beginning of Rua Vieira dos Santos, on the slope of Morro do Pinto, and is a source of drinking water that has supplied the population for many years. Its first tank was built in 1858 by Captain Manoel Pereira Liberato. Its measurements are: 2.60 m long, 1.40 m high; with two iron spouts and paving on the outside (GUARATUBA, PREFEITURA MUNICIPAL, 2009). In 2008, it was revitalized, all paved, the box was renovated, two portals were built, one at the fountain and the other at the entrance, and an image of Saint Louis, the patron saint of the municipality, was placed next to the fountain (GUARATUBA, PREFEITURA MUNICIPAL, 2009). The square was revitalized again in 2010, when the municipality's first open-air gym was installed. (GUARATUBA, CITY HALL, 2010)

FIGURE 39 - LARGO DO CARIOCA SOURCE: The Author (2010).

1.1.1.10 Cabaraquara

This town is home to a large area of Atlantic Rainforest, rivers, mangroves, mountains and a welcoming community that offers many options for visitors, such as hiking trails, handicrafts and, above all, oyster tasting, grown and served in rustic surroundings, in total harmony with nature (GUARATUBA, PREFEITURA MUNICIPAL, 2009). This is where you'll find Morro do Cabaraquara (Cabaraquara Hill), 455 meters high, and a trail that takes you to the top of the hill in two hours, offering a privileged view of the Atlantic Ocean, Serra do Mar and the entire city (GUARATUBA, MUNICIPAL PREFEITURA, 2010).

FIGURE 40 - CABARAQUARA

SOURCE: GUARATUBA, CITY HALL (2010)

1.1.1.11 Cubatao River

According to Guaratuba City Hall (2010), the river rises in the municipality of Sao José dos Pinhais under the name of Sao Joao, serves in part as a border between Sao José dos Pinhais and Guaratuba, receives Arraial and then takes the name of Cubatao until it flows into Guaratuba Bay. Its tributaries are: Castelhano, Coatis, Ribeirao Grande, Zoada, Navio Arraial, Cubataozinho and Rio Preto. It stretches for about 80 km and is navigable by small boats up to Très Barras, and from there upwards by canoes to Salto. During the summer it is used by visitors for swimming and practicing buoy-crossing (PDDI, GUARATUBA, 2002).

FIGURE 41 - CUBATAO RIVER

SOURCE: GUARATUBA, CITY HALL (2010)

1.1.1.12 Barra do Sai

Located on the border between Guaratuba and the city of Itapoà - SC, it has several second homes, a fishing community and little infrastructure (PDDI GUARATUBA, 2002).

FIGURE 42 - BARRA DO SAi

SOURCE: GUARATAUBA, CITY HALL (2010)

1.1.1.13 Islands

Guaratuba is a municipality with several islands, the main ones, according to data from Guaratuba, Prefeitura Municipal (2010): Ilha Itacolomi, Ilha do Sai, Ilha da Sepultura, Ilha da Pescaria, Ilhas do Capim and Ilha do Rato.

1.1.1.14 Mother Church

According to Mafra (1952), the Igreja Matriz Nossa Senhora do Bom Sucesso began to be built in May 1768 and received its blessing on April 28, 1771, from the Reverend Bento Gonçalves Cordeiro, assisted by the friars Joao de Santana Flores and Francisco Borges, who celebrated the first mass on April 29, 1771. Built by the community and designed by Afonso Botelho, it is of colonial religious architecture and has a very simple masonry façade, like other coastal churches of the time. Inside, it is decorated with a discreet altarpiece, probably from the last century (MAFRA, 1952).

According to Guaratuba City Hall (2009), on February 4, 1941, lightning struck the cross on the church tower, which was restored on March 13 of the same year. According to the aforementioned source, the original image of Our Lady of Good Success was stolen in April 1974 and a new image, donated by journalist Cândido Chagas, was enthroned on July 23, 1995. It is a National Artistic Heritage property (GUARATUBA, PREFEITURA MUNICIPAL, 2009).

FIGURE 43 - OUR LADY OF GOOD SUCCESS PARISH CHURCH

SOURCE: The author (2010)

1.1.1.15 Praça dos Namorados and Casarao do Porto

According to Guaratuba City Hall (2009), Praça dos Namorados is located on "Rua da Praia", as it was called when the city was founded (1771), which was later renamed Rua Cel. Afonso Botelho de Sao Payo e Souza, the city's founder, and is on the corner of Rua Prof. Gratulino de Freitas, an illustrious teacher who taught in 1949 in the colonial house, using the light of candles to teach. The square was revitalized in 2009 and is currently used by residents and visitors (GUARATUBA, PREFEITURA MUNICIPAL, 2010).

According to data provided by the Guaratuba Directorate of Culture (2010), the Casarao do Porto is privately owned and there is no documentation on the property, one of the last remnants of the colonial period, although in iconographic terms it is recorded in a watercolor by Debret, from 1827, which allows us to place its construction between the end of the 18th century and the first decades of the 19th century. It has a square plan, built of mixed masonry, stone and bricks, with a four-sided roof, canal tiles, topped by a railing (GUARATUBA CULTURE DIRECTORATE, 2010). It is undoubtedly a significant example of the lineage of townhouses built in the colonial era throughout the country (GUARATUBA, PREFEITURA MUNICIPAL, 2010). The property was used for commerce and housing and was abandoned in the mid-1970s, deteriorating rapidly (GUARATUBA, PREFEITURA MUNICIPAL, 2010). In 1994 it was restored and adapted into a restaurant, and is currently used as Guaratuba's House of Culture (GUARATUBA, PREFEITURA MUNICIPAL, 2010).

FIGURE 44 - CASARAO DO PORTO

SOURCE: The author (2010)

1.1.1.16 Events

The town holds various events throughout the year, especially during the summer season, such as beach events, musical events, music events, etc. The most important and traditional events in Guaratuba are the Festa do Divino and Carnival, according to data provided by Guaratuba, Prefeitura Municipal (2010), as described below.

Feast of the Divine: in the church, opposite the Altar of São Luiz, there is the Altar of the Divine Holy Spirit, whose image was found by a devotee who took it to be bathed in the Itororó Fountain (Largo Nossa Senhora de Lourdes) and then brought it back to the church. As a result, the fountain came to have healing value, which devotees still worship today.

The Divine Holy Spirit is celebrated in Guaratuba every year in the second half of July. The festival, which is attended by a large number of faithful, is announced beforehand by the White and Red Flags of the Holy Trinity and the Holy Spirit, which first go out to the villages, each carrying four revelers. The band is made up of the mestre, the contra, the rabequista and the triple, with instruments such as the viola, the drum and the rabeca.

The Flags began their pilgrimage through distant villages and hamlets, receiving donations to cover the costs of the festival, on a date that was set for May 3 each year, the day of the Holy Cross.

The approach is announced by the symbolic beating of the drum, with the sound characteristic of its function, and then, as she approaches the house, the singing. The owner of the house then goes to meet her, receives her from her conductor and takes her inside the house where, as the Folioes arrive, they sing the verse begun by the mestre, accompanied by the Contra and the *Triple,* with the music of their instruments.

It's a 10-day festival that offers visitors spaces with various types of products, musical and dance performances and a gastronomic fair. The event takes place in Praça Coronel Alexandre Mafra, in front of the Mother Church. In addition to what is on offer, devotees take part in daily prayers and novenas.

The municipality also promotes many other events, such as the "Passion of Christ" and other religious festivals, as well as many sporting events, among others.

Another important event in Guaratuba is Carnival, which is one of the most traditional in Paranâ. There are 5 nights of revelry in the streets of the city, where people have the opportunity to follow the electric trios and the famous Guaratubanda (Guaratuba Band). Every year, the town welcomes around 300,000 revellers.

From the attractions listed above, we can see the great tourist potential of the municipality.

3.3.8 Analysis of tourism support equipment and services

According to data provided by IPARDES (2010), Guaratuba has 147 establishments related to accommodation, food, maintenance, repair, broadcasting and television. This data doesn't just list what matters, it also lists other services. A consultation with the municipality's Department of Tourism (2010) reveals that Guaratuba has 13 hotels, 2 motels, 17 inns, 5 lodges for excursions, 4 *campsites* and 7 associations, totaling 48 lodging establishments, which work at maximum capacity for 2 months during the season and are idle at other times of the year. Of the accommodation establishments mentioned, no hotel closes outside the summer season, but they all work with a smaller number of employees and many of them close some wings and work with just a few apartments. The motels remain open all year round. As for the inns, 5 of them close at the beginning of May and only open for excursions until September. The *campsites* remain open, but demand, according to Guaratuba's Director of Tourism, "is almost zero". According to data from the Tourism Department (2010), two more inns are in the process of being built.

In a consultation with the Commercial Association of Guaratuba (2010), the president argues that in the low season, lodging establishments are very idle and costly, as the maintenance of this equipment becomes high. The president says that the occupancy rate of lodging establishments reaches almost 100% in the high season, in the low season (March, April, July, September, October and November) it reaches 30% and in May, June and August, occupancy doesn't reach 10%. Many entrepreneurs in the hotel business are looking to the Association to promote training and qualification of the workforce and to help the municipality promote events to attract more visitors at times of the year other than the summer, argues the president.

Guaratuba has 23 restaurants of tourist importance, according to the Director of Tourism, and all of them work with lunch and dinner during the season. After Carnival, only 10 restaurants work all year round with both meals, the rest work only with lunch. Of these, 3 close in May and only reopen in September, and another 3 only open on weekends and public holidays. In consultation with the Guaratuba Trade Association, the president argues that the owners of food establishments complain about the lack of qualified labor and say that it is very difficult to keep such establishments open out of season, because the demand in the city out of season is too small to maintain the structure needed for the season.

According to Guaratuba City Hall (2010), the municipality has three ground transportation companies; all of them work with receptive services during the season (very little), but the main demand is students who travel to other cities in search of further study and qualification during the school year. "As most visitors to Guaratuba have their own transport or travel by public bus and/or taxi, these services are rarely used during the high season," said Mr. Nelson, owner of a people transport company.

Guaratuba has 4 companies that provide water transportation, according to Guaratuba, Prefeitura Municipal (2010), but these only work during the summer season or on scheduled tours. Also according to Guaratuba City Hall, the municipality also has 2 tourist agencies, one of which works with inbound services during the high season and outbound services during the rest of the year, but according to the owner, there is very little traffic out of season. The other only works with outbound travel, mainly shopping trips and selling air and road packages.

This data shows that seasonality affects the majority of establishments involved in tourism. In Guaratuba, the drop in demand greatly affects small and medium-sized businesses, generating a multiplier effect, i.e. when there is demand, there are jobs, income, activities, structure, etc. On the other hand, when there is no demand, there is a lack of jobs, income, activities, structure, etc. for citizens.

3.3.9 Analysis of Guaratuba's tourist demand (summer)

According to Guaratuba City Hall (2010), the average floating population in the municipality over the last five years, during the summer season, is approximately 350,000 people and according to IBGE (2009) the fixed population of the municipality is 32,806 inhabitants. A comparison of these figures shows that Guaratuba's population increases more than tenfold during the high season.

The data on tourist demand presented below was provided by the State Secretariat for Tourism in a study carried out on the Paranâ Coast between 2000 and 2006 and published in 2008 in a document called Tourist Region: Paranâ Coast in Data, and analyzed by the author.

According to SETU - PR, the largest source of tourists to Guaratuba is Curitiba, with an average of 53% of those interviewed, followed by other regions of the state of Paraná, with

32%, and the majority of them male. Foreigners represent 2.0% of the total (SETU, 2008).

The car is the means of transportation most used by tourists visiting Guaratuba in all the years of this study, reaching 91.7% in 2002 (SETU, 2008). Own homes are the most used means of accommodation, especially in 2006 with 41.5% (SETU, 2008). The average age of tourists in Guaratuba in 2006 was 39.7 years - a similar average in all the years of the survey (SETU, 2008). The average stay is also very similar in all the years, around 8 days (SETU, 2008). Average daily spending increased from US$ 12.90 in 2000 to US$ 23.10 in 2006 (SETU, 2008). The majority of those interviewed say that this is not their first time visiting the town (92.3%) and that they travel to Guaratuba as a family (63%) (SETU, 2008).

The infrastructure item with the best rating in 2006 was urban commerce, with 82.9% rating it good (SETU, 2008). Also noteworthy was the access infrastructure, which saw a 74.3% increase in ratings between 2004 and 2006 (SETU, 2008). And the item most complained about by those interviewed was basic sanitation and the service provided in support and commercial establishments (SETU, 2008).

Based on the analysis of demand, the very specific profile of those interviewed can be traced, i.e. parents, who travel in their cars and stay in their own homes (characterizing a second home), upper middle class people who visit the coast during school vacations, holidays and weekends (pleasant climate). This analysis confirms the profile of the tourist, which indicates the high seasonality in the municipality.

It is pertinent to state that the above data confirms part of the municipality's tourism potential, the object of this study, which sought to characterize and prove the existence of seasonality in the municipality of Guaratuba, a research that can be observed in the following chapter.

CHAPTER IV - METHODOLOGY

This chapter first describes the work methodology undertaken and then presents the research that sought to characterize and confirm the existence of significant seasonal differences in the municipality of Guaratuba and then comment on their causes and effects, with subsequent analysis.

The empirical study presented in this paper and its results aim to illustrate and contribute to this argument, using Guaratuba - PR, Brazil, as a case study.

4.1 METHODOLOGY

In preparing and constructing the research project for this dissertation, we worked on two interconnected dimensions, as discussed by Severino (1996): the technical dimension, which deals with the scientific rules defining an object, how to approach it and how to choose the most appropriate instruments for research, and the ideological dimension related to a researcher's choices about a topic, what to research, which theoretical basis to use and how to research. Therefore, the personal choice of a researcher is relevant, and as much as impartiality is sought in research, it is believed that scientific knowledge is always culturally and socially conditioned to the historical moment of those who undertake the observation. Based on the words of Severino (1996), it can be said that the object of this research is the seasonality of tourism in the municipality of Guaratuba, located on the Paranà coast, and all the issues related to its development.

The development of tourism in the municipality of Guaratuba, on the coast of the state of Paranà, is well-established, attracting a large number of tourists, especially during the summer vacation months. But what about the other months of the year, how do tourism and other economic activities behave in the municipality? Does the seasonality of tourism in Guaratuba have an impact on the quality of tourist services and on the life of the community in general? Based on the above problem and the need to study seasonal tourism in the municipality of Guaratuba, the following hypotheses were formulated: does seasonality occur due to a lack of tourism planning or due to climatic and territorial factors? Is seasonality due to a lack of commitment on the part of the business community to create reasons (events, programs, new attractions) or a lack of infrastructure? Or does seasonality lead to a drop in the population's income, causing economic and social problems?

As for the nature of the research, this study is original, because as Ruaro specifies:

> Research is considered original when it is a study carried out for the first time, i.e. the focus, the delimitation of the topic has not yet been approached from the aspect being presented. Thus, original research should contribute new discoveries to the expansion of scientific knowledge (RUARO, 2004, p. 24).

In other words, the research carried out in Guaratuba, in its approach to seasonality, took place for the first time. In terms of objectives, this research can be considered exploratory and descriptive, because according to Gil:

> The main purpose of exploratory research is to develop, clarify and modify concepts and ideas, with a view to formulating more precise problems and researchable hypotheses for subsequent studies. Of all the types of research, they are the least rigidly planned. They usually involve bibliographical and documentary surveys, non-standardized interviews and case studies. Sampling procedures and quantitative data collection techniques are not usually applied in these surveys (GIL, 2008, p. 27).

The main objective of this study was to analyze seasonal tourism in the municipality of Guaratuba - PR and its effects on the local community, in order to characterize and prove the existence of seasonality. The specific objectives were: to research bibliographic and documentary data on the proposed topic; to analyze the municipality in question geographically, socially, touristically and economically in order to characterize it; to identify and diagnose the municipality's strengths and weaknesses in relation to its seasonal and annual tourist supply and demand; to analyze the municipality's tourist demand; to identify important territorial elements for the study of tourism and local geography; to characterize and prove the existence of seasonality.

According to Dencker (1998), this type of research is suitable for the phases of literature review, problem formulation, hypothesis development, identification and operationalization of variables.

According to Gil, this research is descriptive because:

> The main objective of research of this type is to describe the characteristics of a given population or phenomenon or to establish relationships between variables. There are countless studies that can be classified under this heading and one of their most significant characteristics is the use of standardized data collection techniques (GIL, 2008, p. 28).

Dencker (1998) argues that descriptive research is indicated to guide the form of data collection when the aim is to describe certain phenomena.

The research was carried out through bibliographic and documentary research and intensive direct observation, which according to Marconi and Lakatos (2008, p. 75) "is carried out using two techniques: observation and interview".

The bibliographic and documentary research was carried out through indirect documentation, because according to Marconi and Lakatos "it is the phase of the research carried out with the intention of collecting previous information, it is done in two ways: documentary research (or primary sources) and bibliographic research (or secondary sources)." (2008, p. 48). This study used documents and publications from bodies such as the Instituto Paranaense de Desenvolvimento - IPARDES, the Instituto Brasileiro de Geografia e Estatistica, the Ministério do Turismo and the Prefeitura Municipal de Guaratuba, among others, characterizing the research as documentary and bibliographic, because according to Marconi and Lakatos "it covers all bibliography already made public in relation to the subject of study, from single publications, bulletins, newspapers, magazines, research [...]". (2008, p. 57). For Manzo[57] (1971, p. 32) quoted by Marconi and Lakatos:

> The relevant bibliography offers the means to define and solve not only known problems, but also to explore new areas where the problems are not sufficiently crystallized, and aims to allow scientists to make parallel efforts in the analysis of their research or manipulation of their information (MARCONI; LAKATOS, 2008, p. 57).

Based on the bibliography analyzed, the appropriate comparisons were made with what is seen in the field, that is, using field research, because according to Marconi and Lakatos (2008, p. 69), "it is that used with the objective of obtaining information and/or knowledge of a problem for which an answer is sought [...]". The same authors also mention that "it consists of observing facts and phenomena as they occur spontaneously, collecting data about them and recording variables that are presumed to be relevant, in order to analyze them." It can therefore be said that this research was also characterized as direct documentation, which "generally consists of collecting data in the very place where the phenomena occur." (MARCONI; LAKATOS, 2008, p. 69).

Within the empirical part, this research is characterized as Intensive Direct Observation, through the two techniques used: observation and interview. The type of observation used in this research was asystematic observation, i.e. according to Marconi and Lakatos (2008) also called unstructured, spontaneous, informal, simple, free, occasional, which consists of collecting and recording the facts of a reality without the researcher using special technical means or having to ask direct questions. It is most commonly used in exploratory studies and does not have previously elaborated planning and control. The interview used was a focused interview, which according to Gil (2008, p.112) "is just as free as the informal interview, but focuses on a very specific topic. The interviewer allows the interviewee to speak freely on the subject, but when the interviewee deviates from the original topic, they make an effort to return to it." The type of sampling chosen was intentional non-probability, which according to Marconi and Lakatos (2008), is the most common type of non-probability sample, and in which the researcher is interested in the opinion of certain elements of the population, but not representative of it.

> This would be the case, for example, if you wanted to know how the opinion leaders in a particular community think. The researcher is therefore not addressing the "masses", i.e. elements that are representative of the population in general, but rather those who, according to the researcher, by virtue of their function, position held or social prestige, are opinion leaders (LAKATOS; MARCONI, 2008, p. 38).

In the case of this research, the sample would be made up of 30 people directly or indirectly involved in local tourism. Of the 30 people contacted, only 11 were willing to answer the interview.

---

57 MANZO, A. J. **Manual para la preparación de monografias:** uma guia para presentar informes y tesis. 2. ed. Buenos Aires: Humanitas, 1973.

In this research, the author observed the behavior of tourists in high and low season and also the environmental impacts caused by mass tourism during the summer season, holidays and weekends.

With regard to the interviews, 11 (eleven) individuals were questioned (one from the food trade, one from the general trade, one second-home tourist, one holidaymaker, one native resident, one from the Town Hall (administrative), one from the Social Welfare, Labor and Citizenship Department, one from the Tourism Department, one from the Environment Department, one from the Civil Police Station and one from the Workers' Agency). These individuals were chosen because of the extent to which tourism in the municipality affects their lives and/or functions.

At the start of the interview, the author gave a brief explanation of her study and the meaning of the term seasonality, then asked the following questions, the answers to which were recorded and transcribed for later analysis: Do you think there is seasonality in tourism in Guaratuba? If so, how does it affect your business and/or function? In your opinion, what are the causes of seasonality in the municipality of Guaratuba? What are the effects of seasonality in Guaratuba? The interviews and observations were carried out in September 2009 and March 2010. These questions were chosen in order to prove the existence of seasonality in the municipality of Guaratuba and its effects on those interviewed.

As for the method of approach, this research can be characterized as hypothetical-deductive, because according to Gil:

> When the knowledge available on a given subject is insufficient to explain a phenomenon, a problem arises. To try to explain the difficulty expressed in the problem, conjectures or hypotheses are formulated. From the hypotheses formulated, consequences are deduced which must be tested or falsified (GIL, 2008, p. 12).

With regard to the method of procedure, this study is characterized as a monographic or case study method, as it "starts from the principle that the study of one case in depth can be considered representative of many others or even of all similar cases. These cases can be individuals, institutions, groups, communities, etc." (GIL, 2008, p. 18).

Finally, the answers obtained were interpreted and analyzed so that the seasonality of tourism could be characterized and proven in the municipality of Guaratuba.

The field research will be presented next, where the data collected will be analyzed.

4.2 FIELD RESEARCH

The field research was carried out by means of a focused interview with an intentional non-probabilistic sample, with 11 (eleven) individuals directly or indirectly involved in tourism and affected by the seasonality of the municipality. The initial idea was to interview 30 people, who were contacted for this question, of whom only 11 agreed to answer the question. We also used systematic observation to analyze factors directly linked to the causes and effects of seasonality. The results of the survey were presented in a table made up of the main causes and effects identified by the interviewees and the results were analyzed in a second stage. The observation was divided into factors for a better understanding, and the considerations are also included in the analysis of the results.

4.2.1 Focused Interview

The focused interview is considered free and, according to Gil (2008), focuses on a very specific topic, in this case, seasonality. Gil (2008) argues that the interviewee should speak freely on the subject, but when they deviate from the original topic, the interviewer should make an effort to get them back on track.

The interviews were conducted in Guaratuba with 11 (eleven) people directly or indirectly linked to tourism: one person from the food trade, one from the general trade, one second-home tourist, one summer tourist, one native resident (fisherman), one person from the City Hall (administrative), one person from the Department of Social Welfare, Labor and Citizenship, one person from the Department of Tourism, one person from the Department of the Environment, one person from the Civil Police Station and one person from the Labor Agency, who analyzed according to their points of view the seasonality existing in the municipality. The interviews and observations took place between September 2009 and March 2010. The main idea behind this interview was to analyze how people linked to tourism in Guaratuba understood seasonality and were affected by it.

At the start of the interview, questions were asked about the subject so that people could understand it better. Figure 2 - causes and effects of seasonality, adapted from Mota (2001, p. 99), which appears in Chapter II, p. 62 of this work, was also shown so that most people would have a better idea of the causes and effects of seasonality. After the appropriate explanations, the questions were asked and the person simply had to speak. All the answers were recorded and transcribed into text, as follows. Based on the interviewees' responses, the author has drawn up a table that describes these considerations in a succinct and organized way for later analysis of the results.

Below are the interviews conducted with a tape recorder and transcribed in full for later analysis.

Interview 1 - V.B. Owner of a restaurant (Churrascaria) in the center.

Do you think that tourism in Guaratuba is seasonal? If so, how does it affect your business and/or function? In your opinion, what are the causes of seasonality in the municipality of Guaratuba? What are the effects of seasonality in Guaratuba?

Answer: Of course there is such a thing as seasonality, we only see people here in Guaratuba in the summer and on vacations, and there are also a few more on July vacation. Seasonality affects my business a lot, because I have to keep at least 20 employees all year round, which doubles during the season, and sometimes they spend a lot of the time doing nothing, so I send them out to clean, wash walls, sand pots and pans, these things, because since I'm paying them, I give them a job to do, and also if an excursion suddenly comes along or there's a good influx of people from here, I have to be prepared. And since they're trained, I have to keep them, because it's very difficult to find "good" labor in this city, so I prefer to keep my own, which is then guaranteed. "To" tell you the truth, we have to stay open to make ends meet, because we get by with the money from the season and another holiday or sunny Sunday, it's hard, but at least 6 months I work in the red. Now you ask me the causes, well, I think the main causes are the school vacations and the two big holidays -

New Year's Eve and Carnival, that's when people have time to travel and I think they also have money, because they spend it well. Think about it, do you think it's fun to come to the beach in the winter, I think it's awful here, there's nothing to do, a lot of beggars and unoccupied people drinking and doing drugs, who would want to come here, not to mention that practically everything is closed. Even those who have apartments or houses come until March and then don't return until September. I still think that Guaratuba is a summer town and I don't think it's going to change any time soon, unless they hold big events and festivals, even for senior citizens it would be "cool", but the health here "is" awful, it would have to improve a lot. Regarding the impacts, I think the biggest is social, most people are unemployed, with nothing "to do", most without unemployment insurance, because they work informally, imagine all these people without money, they get up to a lot of mischief out there, then crime grows, drug use, drug trafficking and many other problems that come with the lack of jobs. Another thing is that the people who live here make very little money in the municipality, so we only see the face of the "money" when the tourists are here. Another thing I notice is that tourists complain about the price, they want the price here to be the same as in Curitiba. That's impossible, we're so far away from our suppliers that it's hard to keep the price down, not to mention the fact that we need to make money during the season, which is why I also raise my prices, because then (in winter) I have to put everything "down" to keep my merchandise turning over. Another thing I see during the season, when I go to the beach - rarely - is how "dirty" these people are, they make a mess of the beaches and the streets, they don't care, they just want to enjoy, party and "party", they don't even worry about the garbage, then they complain "to" us here, "How dirty Guaratuba "is"!", "You don't take care of your city!", this is complicated, they want everything to be beautiful and wonderful, but they don't do anything to help.

Interview 2 - M.C. Owner of a clothing store in the center of Guaratuba.

Do you think that tourism in Guaratuba is seasonal? If so, how does it affect your business and/or function? In your opinion, what are the causes of seasonality in the municipality of Guaratuba? What are the effects of seasonality in Guaratuba?

Answer: Yes, I think there is seasonality and it's quite big, with very high peaks in the summer and idleness in the winter. Well, my business is very affected by seasonality, because during the season I have to keep 5 employees and we still can't handle so many people, then Carnival comes and I have to lay off 4 and I'm left with just one employee, who stays so I can leave, otherwise I wouldn't have to. In January and February, the store is full all day and a lot of money comes in, but from April onwards, there are days when I don't even open the till, I'd prefer it if there was always a little bit, these ups and downs destabilize us, it becomes difficult to invest, improve, grow. Well, as for the causes, I think the main one is the weather, because people hardly ever come to the beach on cold, rainy days (and it never stops raining here). Another factor is school vacations, both summer and winter, because parents can only travel when their children are on vacation and Guaratuba is a second-home town, most people coming here to enjoy their houses and apartments, which are closed all year round. Another thing is that Guaratuba has no options, it's just the beach. If it had events and activities in the off-season, I think more people would come here more often. I also think that there is no tourism *marketing* for the off-season, people don't even know that Guaratuba exists outside the summer season. Look at Balneàrio Camboriù, it has everything, people don't even go there for the beach anymore, they go to enjoy the city, the leisure options, that's what's missing in Guaratuba - leisure options. As for the effects, I think that seasonality brings economic and social problems, because people are without money and jobs for a good part of the year. Crime increases a lot in winter, since during the season there are police everywhere and then Guaratuba becomes a "no man's land". Another thing I've noticed is the price of food, in the season it's almost impossible to eat out, the prices are very high, even the supermarkets take advantage and put everything up, it's absurd. Sometimes it feels like I'm losing my identity as a Guaratuban in order to adapt to the way tourists do things.

Interview 3 - J. G. Owner of a second home in Guaratuba.

Do you think that tourism in Guaratuba is seasonal? If so, how does it affect you? In your opinion, what are the causes of seasonality in Guaratuba? What are the effects of seasonality in Guaratuba?

Answer: Obviously there is, a city with 30,000 inhabitants, receiving more than 300,000 in just two months already characterizes seasonality. Well, I like the summer, but all those people stress me out, the prices get higher, the traffic is unbearable, there are queues for everything, there's a lot of pollution, both environmental, noise and visual. As I have an apartment here and live in Curitiba, I come practically every sunny weekend and on holidays, I love Guaratuba without the fervor, I come to rest, enjoy the scenery, breathe fresh air. I think that Guaratuba, because of its pleasant climate, would be the ideal city for second homes for the elderly, but it's a pity that the health system is still poor, these people could make the economy grow during the winter period. I think that if the City Council invested in a structure to bring these people in, Guaratuba would be a great city for second homes in the off-season too. As for the causes, I think the main one is the lack of infrastructure and leisure activities to attract people all year round. Other causes would be the climate and the school vacation season (most people prefer to travel in the summer and with their children). Another important factor is income: at the end of the year, people receive their thirteenth salary and paid vacations, which helps them to leave home in search of leisure. As for the effects of seasonality, for me the worst thing is the summer crowds: you can hardly leave the house, there are queues, traffic, high prices, etc. But the good thing is that there are so many options for leisure and entertainment. As I come to Guaratuba all year round, I notice that in winter there are a lot of idle people and a lot of crime, perhaps due to the lack of police to cover such a large municipality. Oh, another thing I don't like to see, and which is due to the large influx of people, is the garbage and dirt, visitors don't take care of the environment, I think it's absurd. I'd also like to make a complaint: most places offer very poor service to visitors, and many people only come back if they are well looked after. I believe that the municipality and the business community should invest in training the workforce in the service sector directly linked to tourism.

Interview 4 - J.S. Summer tourist, who stayed in the city for 7 days and stayed in a hostel with his family.

Do you think that tourism in Guaratuba is seasonal? If so, how does it affect you? In your opinion, what are the causes of seasonality in Guaratuba? What are the effects of seasonality in Guaratuba?

Answer: Look, I can't say whether it exists or not, as I've never been to Guaratuba at any other time of the year. But I believe it must exist, because I've heard several locals say that in winter it's very difficult to keep businesses open, that the town is empty and that few people live well here. As far as I'm concerned, the problems that come with seasonality don't affect me, because I'm only here for a week and I enjoy everything the town has to offer, and I can't imagine this place being empty and without leisure options. In my layman's opinion, I believe that the main cause of seasonality here in Guaratuba is the vacation period and I also think that the climate can influence people, because I can't imagine a beach with cold weather. And in my case, the only reason I don't come to Guaratuba more often during the year is because I live far away and need time to stay here and enjoy my rest, and also because I only get vacation time from my job at the end of the year. As for the effects, I think the biggest will be unemployment and the lack of money for the population and, for me, I think it will be the high price of goods in supermarkets and restaurants.

Interview 5 - City Hall - Member of the Administration.

Do you think that tourism in Guaratuba is seasonal? If so, how does it affect your business and/or job? In your opinion, what are the causes of seasonality in the municipality of Guaratuba? What are the effects of seasonality in Guaratuba?

Answer: Yes, there is certainly a seasonality to tourism in Guaratuba. The administration is very affected by seasonality, because we have to manage on state and federal funds for 33,000 inhabitants and receive more than 300,000 people in the summer. You see, the structure to maintain the city is simple, but to cater for visitors requires something more complex in terms of supplies and health. It's very complicated to plan Guaratuba, because residents demand construction work in the neighborhoods where they live and visitors demand structure and activities in the central and coastal regions. You can imagine how difficult it is for public administration to manage all this, I need a lot more people working in the town hall, to meet the demand for people in the city, of course the State Government's Summer Operation helps a lot in terms of personnel for activities related to leisure and security, but in terms of inspection and service, we have hired more than 100 people to work at this time, greatly burdening our payroll. We try to do the best for our population, but these peaks and troughs make it difficult to run a more linear public administration based on the residents' priorities. As for the causes, I can say that, in my opinion, the main one would be the weather, then the school holidays, and then the lack of activities outside the summer season, such as events, parties, congresses etc., the lack of adequate health infrastructure to meet the demand of elderly people who could live in Guaratuba, taking advantage of the mild climate and pleasant air (without pollution). Another cause could be the lack of an adequate tourism *marketing* plan focused on niche markets, but a joint plan between the town hall and businessmen, fighting to bring more people here. The effects are clear, unemployment, crime, lack of money in the local economy, this in the low season, and in the high season it would be abusive price increases, garbage, degradation of the natural and urban environment.

Interview 6 - J. R. da S. Native resident and fisherman.

Do you think that tourism in Guaratuba is seasonal? If so, how does it affect your business and/or job? In your opinion, what are the causes of seasonality in the municipality of Guaratuba? What are the effects of seasonality in Guaratuba?

Answer: It does. It affects us a lot, because when there are tourists in town we work harder and sell more. Out of season, life for us is very difficult, you know, there's that time when we have to stop working so that the fish and shrimp can spawn, the closed season. Even though I receive a salary from the government during these months, I have many colleagues who don't receive anything, so they go around drinking, using drugs, spending all the money they've earned during the season. If we had something else to do in the winter, it would be easier to live, but most people are out of work at this time of year. It makes you sad, but what can you do? I think the "cause" of this business that you mentioned is, first of all, the weather, nobody comes here in the cold, then vacations, we have to have vacations to travel. I think another reason is the lack of things to do. Tourists like going out, partying, eating out, shopping, and there's nothing to do here in winter. As for the "effects", I think the main one is unemployment and then money, people run out of money and then what do they "eat" and "wear"? They'll "steal" and "get up to mischief" out there, "to" have something to "eat".

Interview 7 - Municipal Secretary for Social Welfare, Labor and Citizenship.

Do you think that tourism in Guaratuba is seasonal? If so, how does it affect your business and/or job? In your opinion, what are the causes of seasonality in the municipality of Guaratuba? What are the effects of seasonality in Guaratuba?

Answer: Seasonality exists and is very visible in the municipality. Here at work, we have problems all year round. At the beginning of the season, adventurers appear in search of work, and most of the time they can't get a job, so they stay on the streets depending on handouts and help from the community. This past year, we took these people and sent them back to their towns (we paid for the ticket), which was a way of reducing the number of adventurers. Of course, during the season there are more of them, but they manage to stay because the weather is pleasant (so they can sleep on the street) and people (tourists) end up helping to keep them going. A serious problem is that many of these people are troublemakers and addicted to drugs and alcohol, which requires closer work by the secretariat. After the season, the work starts to get a little more complicated. In April and May, people start to show up who need jobs and help to start a new business, people who are desperate for money or a job. In this sense, the town hall tries to bring in workshops and work with the community so that people's self-esteem doesn't drop and they give up. Between the months of May and August is the worst time, as many people are still receiving unemployment benefits. We need to have activities (tourist or not) every year to reduce the idleness of these people and consequently improve their self-esteem. Regarding the causes, I believe that the climate is an important factor, the availability of free time, like vacations, lack of activities that attract people outside the summer season. As for the effects, I believe that the main one is unemployment and, consequently, an increase in crime.

Interview 8 - F. G. (architect) - Municipal Tourism Department.

Do you think that tourism in Guaratuba is seasonal? If so, how does it affect your business and/or job? In your opinion, what are the causes of seasonality in the municipality of Guaratuba? What are the effects of seasonality in Guaratuba?

Guaratuba does have a seasonal tourism. It's a "resort" town (we can also say that it's sometimes a dormitory town) that survives exclusively on the summer seasons, largely due to its physical conformation, that is, with the opening of the access road to the municipality (before, the town was practically isolated and it took almost a month to get to Curitiba, This was reduced to 6 hours with the road and today to 1.5 hours with the BR-101 highway), and the subdivision of land near the edge of Praia Central began, and this phenomenon has been repeated to this day (extending along the access route to Guaratuba, between the Brejatuba and Nereidas neighborhoods). Returning to the question of seasonality, it can be seen in the long period of time between the summer season (formerly from December to March, and today from December to February at the latest) and the winter season (July, when the vacations and the Festa do Divino take place), without forgetting the occasional vacations (Easter, Labor Day, Corpus Christi, Sete de Setembro and Quinze de Novembro). Seasonality is, in my opinion, closely linked not only to the view (held by the local population, tourists and vacationers) of the city as a mere seaside resort that survives on the seasons and the tertiary sector (services and commerce), but also to the mentality (local culture) already assimilated by sporadic work, in a short space of time and aiming for high profits. You think: I didn't win yesterday, I want everything today and nothing tomorrow. There was almost always no medium or long-term planning (city hall-state government-businessmen), there were no differentiated investments or innovative projects, far from the commonplace. This sameness condemns local tourism to eternal seasonality. Without a better hotel infrastructure and proper exploitation of the existing attractions (natural, cultural and historical), Guaratuba will be doomed to receive a public with ever-decreasing purchasing power, driving traditional families from Paraná away to Santa Catarina and the coasts of São Paulo and Rio de Janeiro (if not the Northeast - much exploited and so much more lacking in infrastructure than our Guaratuba). The causes of the seasonal nature of tourism in Guaratuba are the lack of a better infrastructure in general (hotels, gastronomy, promotion of the municipality), the need to greatly improve the support structure for tourism (training local agents and the workforce in general) and, above all, the idea of a small town (backward and out of time) in relation to its enormous territory and potential for tourism development (the subject of endless discussions, but lacking in realization). This mentality needs to be changed in the medium and long term with the implementation of a major economic and tourist development master plan for the region. The effects of the seasonal nature of tourism in Guaratuba are easy to see: supermarkets are short of stock, road traffic and parking spaces can't cope with the demand during the high season, restaurants aren't prepared for large crowds and can't survive periods of low tourist numbers (as well as hotels, shops, services, etc.), public security isn't up to the task.), public security is inefficient, health depends on a regional hospital, jobs revolve around the town hall, construction and tourism (commerce, services, without any different, innovative activity, with appeal and attractiveness to a larger public). There is no cinema, no public library, only one museum (Casa da Cultura), no pavilion or public space with a suitable infrastructure for festivities, fairs and municipal products. Nautical tourism is almost unexploited, not to mention ecological tourism, with parks that only exist by decree and that don't have the slightest support, protection or inspection structure. In this sense, we are insisting on hiring a support team to guarantee the execution of a major sustainable development project for Guaratuba. We are counting on resources from the PAC-CH, partnerships with APPA, Prodetur, among other programs, as well as the return of investments from the state government, which has not talked to Guaratuba for many years due to the sheer lack of interest on the part of previous municipal administrations.

Interview 9 - Municipal Environment Department.

Do you think that tourism in Guaratuba is seasonal? If so, how does it affect your business and/or job? In your opinion, what are the causes of seasonality in the municipality of Guaratuba? What are the effects of seasonality in Guaratuba?

Answer: Tourism is seasonal, and that's a fact. Our department tries to make tourists aware of the importance of preserving the environment, of not dirtying public roads and beaches, of connecting their homes to sewage treatment, but this is very difficult in the summer, because most visitors are only concerned with their summer vacations. But for us environmentalists, it's important that there is a break in the number of visitors, so that the natural environment near the beach can breathe. But Guaratuba doesn't just have beaches, it has an APA and parks that need constant monitoring. There are still some sustainable development projects, involving traditional communities, projects to raise native marine species in confinement, such as oyster and shrimp farming. As for the causes, I think the main one is the climate, because people like to come here in the summer, and I think the end-of-year vacations are the strong point of the season. I don't think people come here in the off-season because it's dead, so what are they going to do here in Guaratuba in the cold? As far as I can see, the main effects are likely to be unemployment and a lack of skilled labor, both in summer and winter.

Interview 10 - Guaratuba Civil Police Station.

Do you think that tourism in Guaratuba is seasonal? If so, how does it affect your business and/or job? In your opinion, what are the causes of seasonality in the municipality of Guaratuba? What are the effects of seasonality in Guaratuba?

Answer: Yes. During the season, we do our work on call, because the state government sends hundreds of military and civil police officers to do their work on patrol and on call. This is a problem, but it's important to say that seasonal crimes are generally linked to traffic, drug use, fights, robberies and homicides are few and far between, and most of these crimes are committed by visitors. The problem is that after the summer season, municipal crime increases, as do burglaries, drug trafficking, fights between friends and family, robberies and murders. To give you an idea, in 2009 we had one homicide during the summer season and 36 homicides between March and August, and the main motive was drug trafficking and use.

One of Guaratuba's biggest problems outside of the summer season is drug trafficking and use, which leads people to commit major crimes. Perhaps unemployment and the lack of economic stability are some of the reasons for this problem. You asked me about the causes of seasonality. I think the main factor is the climate, but the lack of things to do means that people don't want to come here outside of the high season. I think people want to come to the beach on vacation, when they have a lot of free time and a lot of activities to do. I think the effects, as I said before, are an increase in crime and unemployment.

Interview 11 - S. L. da L. Guaratuba Labor Agency.

Do you think that tourism in Guaratuba is seasonal? If so, how does it affect your business and/or job? In your opinion, what are the causes of seasonality in the municipality of Guaratuba? What are the effects of seasonality in Guaratuba?

Answer: Yes, there are. Well, it's very interesting to analyze how the seasonality of tourism directly affects my service. Let's start with the month of September, when second-home residents are looking for people to work as janitors to repair their properties and businesses start looking for labor to train them for the season. Between September and February there is very little demand for unemployment insurance. In November and December there is a *boom* in demand. Everyone is desperate for people to work, in practically every area. The only sector that decreases demand at the end of the year is construction. In January and February there is still demand, but on a smaller scale. In the next three months, March, April and May, we work like crazy applying for unemployment insurance and FGTS, at which time most of the people who start work at the end of the year are unemployed. In June, July and August, we have practically nothing to do. The workload is minimal and the number of applications for loans at Banco Popular (which is attached to the branch) is huge, so much so that at that time I'm sent to Banco Popular to "lend a hand" to the person in charge. Our

my work is very different from other cities that have a stable population. Through my work, I can clearly see the seasonality of service provision in the municipality. Regarding the causes of seasonality for tourism, I think the main one is the climate and secondly the vacation season, most people travel when their children are on vacation. My vacations, for example, are in August, and we rarely travel because my children are at school. I think Guaratuba has good hotels and restaurants and should attract people to enjoy these places out of season too, through events, *concerts*, meetings etc. I think there is a lack of tourism *marketing* to publicize these spaces and attract more people. With regard to the effects, for me the main one is unemployment followed by economic problems faced by the population out of season. Perhaps the increase in crime is also an effect of seasonality.

The analysis of the data collected follows in section 4.3.

In this research, the author carried out systematic observations of natural and institutional seasonality in the municipality of Guaratuba, analyzing mainly climate, socio-economic and socio-cultural factors.

4.2.2 Systematic observation

In order to carry out asymmetric observation, which according to Marconi and Lakatos (2008) is an unstructured technique, also known as spontaneous, simple, free, etc., consists of collecting and recording the facts of reality without the researcher using specific technical means or having to ask direct questions. We took into account the categories of causes of seasonality according to Hartmann[58] (1986 *apud* ALMEIDA; KASTENHOLZ, 2009, p. 5) who subdivides seasonality into natural and institutional.

4.2.2.1 Natural Seasonality in Guaratuba

As mentioned earlier, natural seasonality encompasses several factors directly linked to climate variation.

As far as the climate is concerned, Guaratuba is still very dependent on the sun and the heat. As a seaside resort with virtually no leisure options and infrastructure outside the summer season, it suffers from the seasonal imbalance in demand, which is linked not only to the type of demand, but also to the existing supply.

Tourists visit the town mainly because of the beach and the heat, they don't care about the prices or the traffic jams, what they really want is to enjoy what the town has to offer at this time of year. At other times, characterized as low season, tourists, usually second homes, come to the town at weekends to relax and enjoy the milder climate, since even in winter Guaratuba has a pleasant climate.

The mass tourism that takes place during the high season has a major impact on the environment, as it dirties the streets and beaches, pollutes rivers, seas and the bay (through sewage that is dumped directly into the sea), and degrades preserved areas such as restinga and mangrove swamps.

What we can see is that as long as Guaratuba doesn't offer leisure options that don't depend on the sun and the beach, the scenario won't change. Residents are already conditioned to working hard in the summer and less or almost nothing in the winter, so the municipality is at the mercy of the civil service and the poor quality of the services provided for much of the year.

4.2.2.2 Institutional Seasonality in Guaratuba

Institutional seasonality is closely linked to socio-economic and socio-cultural factors, i.e. it depends on school vacations, professional vacations, public vacations and the national, state

58 HARTEMANN, R. **Tourism, seasonality and social change**. Leisure Studies, 5 (1), p. 25-33, 1986.

and municipal calendar of events, as well as income issues.

In addition to the climate, the institutional seasonality is also well characterized in Guaratuba, since it has practically no distinctive attractions for seasons other than summer. People are conditioned to travel to Guaratuba during school and professional vacations, i.e. generally during the summer months (December to March) or during the winter school vacations (July) to a lesser extent. What is noticeable is that in winter, the majority of visitors are second-home residents who come to the town to spend a few days or weekends in their homes. An important factor that brings these people in during the winter is the Festa do Divino (10-day religious festival), which takes place in July (Guaratuba, Prefeitura Municipal, 2010), making the town bustle during this period.

During the summer, the municipality of Guaratuba, in partnership with the State Government and local businesses, offers a wide range of leisure, sporting and cultural activities, activities that don't take place at other times of the year, because the municipality doesn't have the funds for such investments, the State Government withdraws its "Summer Operation" and most businesses close their doors to avoid the risk of bankruptcy due to lack of demand. As long as there are no different options for activities and/or events at different times of the year, Guaratuba will only remain linked to "sun and beach" tourism.

Guaratuba doesn't need to lose its identity as a seaside resort, where beaches and leisure in the sun are of great importance, but it could offer different tourism options at different times of the year, since it is a municipality with a large area of land, with areas of environmental protection and involvement in cultural tourism, events, ecotourism, nautical tourism and rural tourism. Its vocation is for "sun and beach" tourism, but its infrastructure and equipment could be used for other aspects of tourism at other times of the year.

4.3 ANALYSIS OF RESULTS

In order to better visualize the data collected and then cross-reference and analyze it, a summary has been drawn up (table 5), showing the responses of those interviewed in the field research (focused interview). The first gap in the table shows who the interviewee was and the second gap shows how they were affected by the seasonality. The third gap shows the causes of seasonality, according to the interviewee, and the last gap shows the effects of seasonality, according to the interviewee's opinion.

Interviewee	How it is affected by seasonality	Causes	Effects
Restaurant Owner - Churrascaria (open all year round)	* A lot of manpower is needed during the season and in the off-season	* School vacations and public vacations; * Lack of things to do out	* Unemployment; * Lack of money in the local economy;

	professionals are idle; * He works in the red 6 months of the year. * Lack of skilled labor	of season; * Trade practically closed and without quality; * Lack of health infrastructure; * Climate.	* Increase in crime; * High prices in summer; * Garbage and dirt in season.
Owner of a clothing store (located in the center of Guaratuba)	* Number of employees increases a lot during the season from 1 (off season) to 5 (in season) * High turnover in summer and almost nothing in winter; * Working in the red for much of the year (stock-outs)	* Climate; * Vacation period (summer and winter); * Second home tourists have no leisure options; * Tourism *marketing* is lacking.	* Low-income population in the winter period; * Abusive prices in season; * Increase in crime; * The local population loses its identity.
Second Home Tourist	* High prices in season; * Traffic just like in the big cities; * Queue for everything; * Environmental, noise and visual pollution.	* Lack of health and leisure infrastructure out of season; * Climate; * Vacation period; * Extra income at the end of the year.	* Unemployment; * Poverty in winter; * High prices in season.
Summer tourist	* High prices	* Climate; * Vacation period; * Distance.	* Unemployment; * Little off-season cash turnover; * High prices.
Local Resident - Fisherman	* More income in summer; * The rest of the year there is practically no income; * There's a lack of work in winter.	* Climate; * Vacations; * Lack of things to do.	* Unemployment; * Lack of income; * Increase in crime.
City Hall - Administrative	* Administrative and urban planning; * The supply and health structure; * More people to work in the town hall in the summer; * Difficulty in meeting the priorities of permanent residents in the summer.	* Climate; * Vacation period; * Lack of events, parties, congresses; * Lack of medical facilities; * Lack of a tourism *marketing* plan.	* Unemployment; * Increase in crime; * Low-income local population; * Garbage; * Degradation of the natural and urban environment.
Department of Social	* Social problems in the	* Vacations;	* Low self-esteem among

Welfare, Labor and Citizenship	summer with beggars and begging and in the winter with idle and indebted people;	* Climate; * Lack of activities in winter.	the population during the off-season; * Increased
	* The need to provide assistance to these two audiences at different times and with different solutions.		criminality; * Unemployment.
Tourism Department	* More work; * You've noticed that the summer season has shrunk from 3 to 2 months.	* Lack of general infrastructure and support for tourism; * There is a lack of skilled labor; * The idea of a small, backward town; * Lack of leisure activities; * Lack of adequate infrastructure for events; * Little nautical tourism and ecotourism, * Little security; * Poor health infrastructure; * Lack of a master plan for economic and tourist development.	* Sporadic work; * Unemployment; * Lack of regulatory stock in supermarkets; * Low load capacity in traffic; * High prices; * Lack of skilled labor during the season; * Idleness of services out of season.
Environment Department	* In the summer season you work more with tourists; * In winter he works more on sustainable projects with the local population.	* Climate; * Vacations; * Lack of activities outside the summer season.	* Unemployment; * Lack of skilled labor.
Guaratuba Civil Police	* I work less in the summer because of the state government's Summer Operation; * Problems with open surveys in the summer season; * Out of season they suffer from drug trafficking, break-ins and murders.	* Climate; * Vacations; * Lack of leisure options outside the summer season.	* Increase in crime; * Unemployment.

Guaratuba Labor Agency	* Very seasonal service, with busy times and idle times;	* Climate; * Vacations; * Lack of activities outside the summer season; * Lack of tourism *marketing*.	* Unemployment; * Increase in crime; * Lack of money in commerce; * Indebtedness of the local population.

TABLE 12 - CAUSES AND EFFECTS OF SEASONALITY IN THE MUNICIPALITY OF GUARATUBA.
SOURCE: Prepared by the author (2010)

By analyzing the data obtained through observation and focused interviews, it is possible to characterize the seasonality of tourism in the municipality of Guaratuba. In addition to the data collected, data provided by the city council, such as the number of floating population and the SETU - PR demand study, were essential for this analysis.

From the study above, it can be said that seasonality affects most of the people interviewed. In the retail sector, owners complain negatively when it comes to the number of employees, both in terms of quantity and quality, because there is a big problem with turnover, since they can't keep many employees outside of the summer season and the new hires are generally not qualified. There is also a major problem with high expenses throughout the year, which means that establishments are running at a *deficit for* much of the year.

With regard to the two types of tourists interviewed, it is said that tourists who come for a few days only complain about the inflation of goods, when compared to their place of origin; while second home tourists, who visit the town periodically, not only complain about inflation, but also say that they are affected by the large flow of cars, the queues in various places and, above all, the pollution, as they visit the town out of the summer season and notice how dirty the town gets during this period.

With regard to the local resident from a humble background, such as the fisherman, the following is analyzed: during the summer he has a higher income and lives better, but during the rest of the year he has practically no income and complains about the lack of work during the winter period. It's important to note that he is an example of a citizen, like many others, who spend much of the year at the mercy of small "odd jobs" and weekend or holiday activities in order to make a more dignified living.

In terms of administration, the town hall doesn't have adequate administrative and urban planning for all times of the year. During the summer, the municipality's infrastructure, especially in the supply and health sectors, is stretched to the limit and the administration is often unprepared. At the same time, the municipality needs to hire many people for temporary jobs, which requires an increase in the budget. Another difficulty encountered by the municipal administration is meeting the priorities of the city's residents during the summer, who are dependent on seasonal structures and activities.

The Department of Social Welfare, Labor and Citizenship works all year round to help people with social and financial problems. In the summer season it needs to minimize the number of beggars and begging on the city streets through programs, and during the other nine months of the year it needs to assist residents who are unemployed, idle, have low self-esteem and no money to pay off their debts. This department has very different roles at different times of the year.

The tourism and environment departments analyze that there is more work during the season with tourists, but that during the year they need activities that minimize the negative impacts of seasonality and more sustainable projects with the local community.

The Guaratuba civil police officer argued that during the summer season the work is better than during the off-season and that during the high season the state government's Summer Operation is prepared for the crimes that occur. As there is a lot of policing during the summer, the offenses aren't so serious, but as soon as the summer ends, the big crimes begin and crime increases, along with the service provided by the municipality's civil police. Drug trafficking and use are the main causes of homicide in the municipality.

At the Workers' Agency, the agent says that there is a lack of linear work, because like any activity in the city, its service is also dependent on the summer season and the demand for work and credit lines, since Banco Popular is attached to the Workers' Agency (strategically).

Through the focused interviews, it was possible to characterize, according to the interviewees, the causes and effects of seasonality. From there, it was necessary to re-read figure 2, which was re-elaborated by the author, according to the causes and effects of seasonality in the municipality of Guaratuba specifically.

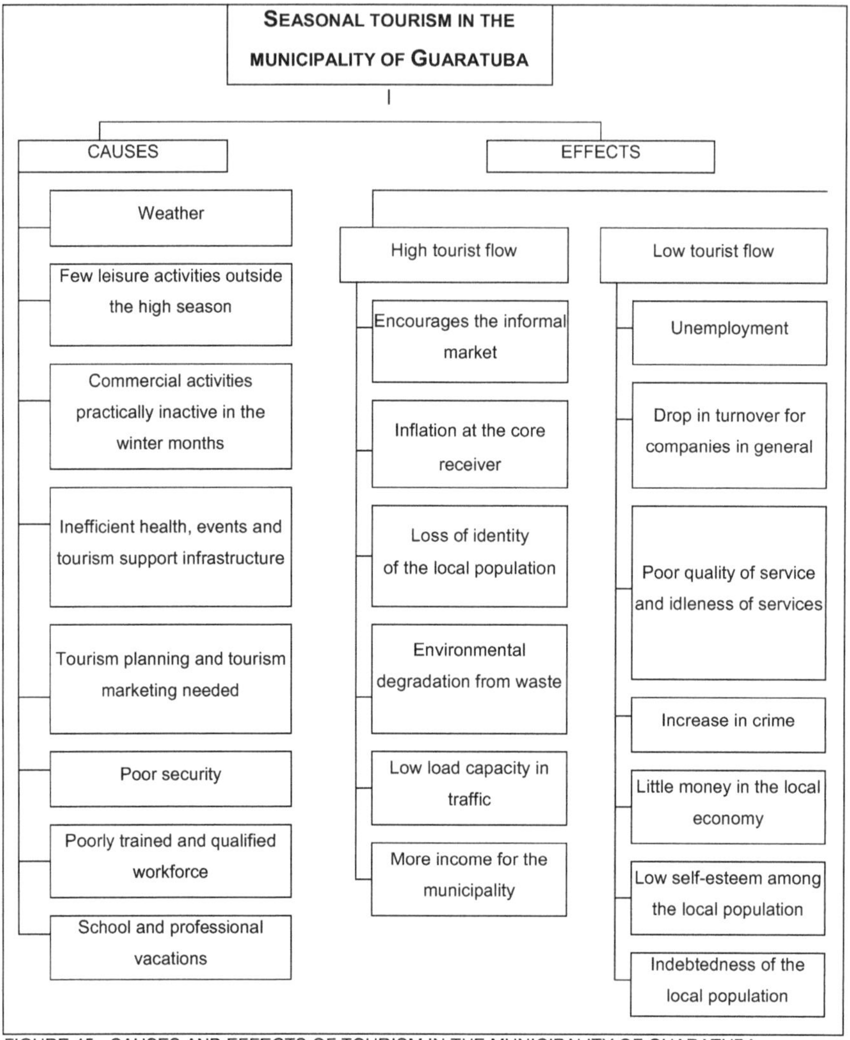

FIGURE 45 - CAUSES AND EFFECTS OF TOURISM IN THE MUNICIPALITY OF GUARATUBA

SOURCE: Prepared by the author (2010)

Figure 45 shows that Guaratuba is characterized as a seasonal city, with its own peculiarities in relation to tourism and which still depends essentially on "sun and beach" tourism and suffers from the causes and effects of seasonality, since the negative effects outweigh the positive ones.

Mota (2001), in creating the figure, describes the causes and effects of seasonality in a "generic" way; in this study we sought to identify the causes and effects of seasonality in the municipality of Guaratuba specifically, and came to the conclusion that the main causes of seasonality in the municipality are: the climate, the lack of leisure activities outside the high season, practically inactive commercial activities in winter, inefficient health, events and tourism support infrastructure, the lack of a tourism *marketing* plan, a poorly trained and qualified workforce, poor security and school and professional vacations. When asked, the majority of respondents said that the weather and the vacations were the main causes of seasonality, but on closer examination, the city could reverse this problem if it provided off-season visitors with more health infrastructure, events and quality services, attracting more visitors at other times of the year.

As for the effects, the high influx of people means that the informal market is encouraged, prices are pushed up and the local population loses its identity as it tries to copy the visitor's culture. The environment is further degraded by the number of people who use it and the traffic that can't cope with the number of cars. The good thing about all this is that the municipality earns more and the locals get extra income at this time. However, when there is a low influx of people, and this happens for most of the year, the effects are more drastic, because there is a big increase in the number of unemployed, companies' turnover drops sharply, there is a drop in the quality of services provided and idleness, crime increases, because without job and income generation people are tempted to go down the path of crime. At this time, people's self-esteem plummets and debt is a certainty.

Analyzing Guaratuba, we can see that the municipality is part of the exclusionary model of tourist development, which is just one of the creations of the archetype of the world's dominant society, which has, on the one hand, a minority enjoying the "benefits of modernity" and, on the other, the majority of the population living in misery or having their survival strategies threatened.

What has been observed is the simple adaptation of various localities to this current model of society, meeting, on the one hand, the interests of a minority benefiting from the profits of the new forms and uses of commercialized territory, and, on the other, the demands for leisure and its consumption patterns imposed by the very agents who produce space, created and exploited by national and international real estate and tourist capital. The inclusion of the population in tourism development will only happen by combating this reality. To do this, it is necessary to build strategies that respond to the current reality in a differentiated way and without demagoguery.

An active stance on the part of civil society is essential. As long as individuals consider themselves to have no significant role to play in history, understanding their actions as only individual, linked to satisfying their personal desires, they will be subject to the exclusionary logic and its adverse consequences.

Guaratuba needs to be placed within the context of medium and long-term planning, with strategies to ensure that its territory is used in a coherent way by municipalities and by the expected demand for people, both in high season (summer) and at other times of the year, looking for ways to attract people who can enjoy what the municipality has to offer.

4.4 General comments on the survey

Guaratuba is a municipality that suffers the effects of seasonality every year. Its current population is around 33,000 and during the summer season it receives around 350,000 people. In order to support this contingent, it needs adequate basic and support infrastructure, good service provision and a qualified workforce.

This study shows that there is a general consensus on the complexity that characterizes the seasonality of tourism, not so much in terms of its consequences, but particularly in terms of its causes. From this it can be seen that the negative effects should be dealt with mainly at the cause level, since what is lacking is a more in-depth knowledge of the explanatory factors of the structural matrix on the part of the receiving municipality and of the behavioral matrix on the part of the visitor. Almeida (2000) argues that the nature of the relationship between seasonality and visitor motivations is not known, and questions such as whether it is dissatisfaction with conditions at the destination or desire for the attractions at the destination that play the biggest role in shaping seasonal tourism patterns are also a mystery. But in Guaratuba the motivations seem to be closely related to the climate and the leisure activities offered by the town during the high season. It is important to consider the development of various efforts to explain not only the causes, but also their interconnections with the motivations and constraints of visiting in different seasons, in order to better understand the phenomenon and consequently formulate more effective strategies for dealing with it.

Diversifying demand is seen as one of the most important strategies in this context. The most effective implementation of this strategy in a destination area directly requires an in-depth understanding of the behavior of different tourist segments (and possibly different origins) at different times of the year.

It can be seen from this that Guaratuba has practically no tourist public in the low season, apart from second residents who visit the town on weekends, holidays and July vacations. In the high season (three months of summer), demand is characterized by upper-middle-class Curitiba residents, the vast majority of whom are also second residents, and by peaks in the movement of people from other origins at the end of the year and during Carnival. This shows that the main motivations for these people to come to Guaratuba are the climate, the ease with which they can travel at these times because of the free time, the extra income and the activities that the town offers its visitors.

It can be concluded that climatic and structural factors attract people to Guaratuba in the summer and these same factors drive them away in the winter, determining the type of seasonality that exists.

From an economic point of view, the high season market may seem more attractive, implying higher revenue per tourist, but it also burdens local spending, generating widespread inflation. In this context, there are also many adventurers and the informal market, which doesn't help the local economy much, but only generates more competition and minimizes the amount of revenue that should stay in the municipality.

In addition, by analyzing the causes of seasonality, the municipal administration together with local entrepreneurs should propose new attractions, products and services that could increase people's interest in visiting Guaratuba and consequently reverse the negative effects of seasonality.

Seasonal markets are especially different from an economic and social point of view, and within their possibilities, as is the case in Guaratuba, they reveal different forms of local and tourist behavior throughout the seasons. In general, any destination seeks to attract tourists all year round, but understanding the differences between seasonal market segments can help to identify, select and better serve each group, thus increasing the potential effectiveness of the strategy of diversifying tourist demand.

Identifying the publics in the municipality and, above all, identifying the causes of the phenomenon can help in more effective forms of planning with more interesting results for the actors involved.

Municipalities that live with the negative effects of seasonality should prove the existence of the phenomenon in all its breadth and analyze ways to minimize its impacts, especially on the local community, and review ways to mitigate it in order to attract more visitors in seasons with fewer tourist flows.

CONCLUSION

This study sought to demonstrate the existence of a general consensus regarding the complexity that characterizes the seasonality of tourism, not so much in terms of its consequences, but particularly in terms of its causes, with the municipality of Guaratuba as a case study.

Tourism is materialized in the logic of the geographical differentiation of places and regions (MOLINA, 2007). It has, in fact, become important to social scientists for understanding what is known as local and regional development and society itself. It concerns space, because the social relationships that have taken place over time alter and restructure spaces produced at different times, building cultural heritage, cities, in other words, the places that become the object of the tourist's gaze.

Cities with their services and facilities, natural and cultural elements are inserted and distributed in different ways, giving tourism a spatial dimension. These places are inhabited and the inhabitants are therefore the hosts. Tourism is an activity that involves the consumption of spaces in a variety of ways, structuring landscapes and businesses and giving agility to processes with a great capacity for territorial organization. Each type of tourism creates a demand for space, whether it's in the mountains, on the coast, in urban or rural areas.

Looking in more detail at the strengths and weaknesses of the municipality of Guaratuba, we can see that it still has buildings of historical and cultural value, as well as ample natural value through its beaches, forests and rivers. Guaratuba is located on the shores of Guaratuba Bay and can be reached from Curitiba by *ferry via* the BR-277 highway, or via the BR-376 highway, via Garuva-SC, making it easy for visitors to reach it by road, approximately 117 km from Curitiba.

Its signage, both tourist and informative, is still very poor. Guaratuba is a municipality with a large territorial area and plenty of space to develop "sun and beach" tourism and events, as well as a large rural and environmentally protected area. It was found that the current government is interested in involving the community in activities, but the community in general is still not very involved in the process of raising awareness of tourism.

A very positive factor to consider is the current political interest in the planning and orderly management of tourism, but even so, Guaratuba still needs to improve in terms of tourism support equipment and infrastructure. This is probably because Guaratuba is a highly seasonal municipality, with little foreign exchange coming in for most of the year, making it difficult for businesses to invest.

Guaratuba has preserved natural attractions with tourist potential and appeal, such as the community of Cabaraquara, where there is still a large patch of Atlantic Forest, rivers and mangroves, mountains and a welcoming community that offers a variety of options for visitors. Trails, handicrafts and, above all, the tasting of oysters, grown and served in rustic surroundings. The trail that leads to the top of Cabaraquara hill offers a beautiful view of the Atlantic Ocean, the Serra do Mar and the whole city.

In addition to Cabaraquara, the municipality has the Guaratuba Environmental Protection Area (APA de Guaratuba), declared a protection area in 1992, with the aim of safeguarding the biological, scenic and cultural aspects of an approximate area of 199,596.50 hectares of Atlantic Forest and associated ecosystems. Its perimeter begins at the Serra do Mar Reservation Line and covers part of the municipalities of Guaratuba, Matinhos, Tijucas do Sul, Sao José dos Pinhais and Morretes. The aim of the Guaratuba APA is to make the rational use of the region's environmental resources and the orderly occupation of the land compatible, protecting the water network, the remnants of the Atlantic Forest and mangroves, archaeological sites and the diversity of fauna and flora. It should be noted that the creation of the APA has disciplined tourist use and guaranteed the quality of life of the caiçara communities and the local population. Another place that seeks to preserve the environment is the Boguaçu State Park, created in 1998, with an area of 6,052 ha, with the aim of ensuring the preservation of the mangrove ecosystems, the restinga around the Boguaçu River and Boguaçu Mirim. In addition to its important preservation function, the Boguaçu State Park could become an urban leisure and cultural facility, aimed at raising community awareness of the importance and notability of the coastal landscape.

It can be seen that the current government is concerned about the quality of the tourist structure, which can be enjoyed by residents and visitors. But it is necessary to look for ways to improve the basic infrastructure (sanitation, streets, sidewalks, public transport), which is still precarious for the amount of demand expected for the high season. Another important factor that must also be taken into account is the health system, which still has serious deficiencies, both in the high and low seasons.

Guaratuba's main attractions are its beaches and Guaratuba Bay, and this entire environment is fundamental to maintaining the life of fish, birds, crustaceans, molluscs and mammals and attracting visitors. The dynamics and cycle of storms and rains distribute nutrients to the beach and the interior of the dunes, giving rise to new shoots of vegetation and shelter for animals. If we don't pay attention to this environment, very little will change and visitors and residents are the main players in this process of visitation and preservation. As well as the beaches, visitors can enjoy the landscapes and visit some of the hills within the municipality, such as Morro do Pinto, Morro Morretes and Morro Espia-Barco. But most tourists visit Guaratuba in search of the beach and look for waters with good bathing conditions and minimum infrastructure to enjoy their vacations, holidays and weekends. Praia Central is the city's main tourist attraction during the summer season and should be improved to please its visitors and attract more.

During the summer season, the municipality's image is shown daily by the media, as a way of spreading the word about its tourist conditions, but on closer inspection, the municipality's image is still deficient (poor) in the emitting nucleus and in possible emitting nuclei throughout the rest of the year, and needs to be improved in order to attract people during other times of the year. Another important point to consider is public safety: Guaratuba has security problems at times when there is little demand from tourists, causing residents and potential visitors to fear violence.

As for accommodation, although Guaratuba has many important places, such as hotels and associations, own houses or rented accommodation are still the main forms of lodging, characterizing the clear existence of second-home tourism.

For the orderly development of tourism, the municipality would need more projects, tax and financial incentives, since the budget is limited for specific tourism policies and tourism *marketing.* Not to mention that the quality of service provision is still poor, due to a lack of trained and qualified professionals.

As for the rural region, the Cubatao River area could be better exploited for tourism if there were more infrastructure and publicity for the area, as there is great potential for ecotourism and adventure tourism.

In short, Guaratuba is a municipality characterized by "sun and beach" tourism and clearly dependent on income from the high season (summer), which causes major social and economic impacts during other periods of the year. Guaratuba, if well planned, could become a hub for tourism in Paraná all year round, trying to combine event tourism with leisure tourism and exploring ecotourism and its cultural, historical and natural heritage, thus enhancing the best the city has to offer and consequently creating new activities that generate jobs and income, whether through commerce, handicrafts, nautical tourism, fishing (sporting or not), observation tourism, among others.

This study set out to characterize seasonality and prove its existence, i.e. times of high demand for space, leisure and infrastructure and times with little or no demand, in the municipality of Guaratuba - PR through a survey of 11 individuals directly or indirectly linked to tourism. The research problem asked whether the seasonality of tourism in Guaratuba had an impact both on the quality of tourist services and on the life of the community in general. It can be concluded from the study that there is a low quality of service provision in the municipality at all times of the year, and that the local community is the one most affected by the negative effects of seasonality, as it is left without qualifications, jobs, debt and crime problems. The hypotheses initially formulated asked the following questions: would seasonality occur due to a lack of tourism planning; would it be due to climatic and territorial factors; would seasonality occur due to a lack of commitment on the part of the business community to create reasons (events, programs, new attractions) or due to a lack of infrastructure? Or would seasonality lead to a drop in the population's income, causing economic and social problems?

When analyzing the results of the survey, it can be seen that there is little specific planning for tourism and that climatic factors are the main cause of seasonality in the municipality, since all the interviewees said that the climate is the main factor that causes seasonality. Regarding the factor related to the reasons for the occurrence of seasonality, it was found that there is a real lack of complementary activities to the natural ones, such as events, leisure, motivating programs for people to visit the town on other occasions than the summer season. Regarding the socio-economic problems arising from seasonality, it is considered that these were the most evident in the field research, since residents in Guaratuba have several competitive disadvantages, such as unpreparedness and low economic power, especially outside the summer season. These findings reinforce the need for tourism planning, integrating the public and private sectors.

The existence of tourism planning carried out by public authorities should serve the population of tourist locations, recognizing the heterogeneity of this population. In this way, it would become more possible to adopt a tourism development model that wasn't just geared towards meeting the desires of tourists (with a view to the economic aspects of the activity) and wasn't based on copying models from outside these locations, brought in by the new agents that began to exploit the territory. According to Molina (2007), the inclusion of the population often comes up against the interests of hegemonic agents (politically and/or economically) who benefit from the dominant logic of tourism development. In this power game, the marginalization of the population is perpetuated and social inequalities are exacerbated.

In order for the negative effects of seasonality to be at least minimized, it would be necessary to regulate the public-private relationship, in which all the actors involved could contribute to the tourist development of the municipality, based on existing coastal environmental management legislation, and the local actors, the main drivers of the application of integrated management programs, could guarantee the lasting development of coastal regions such as Guaratuba, whether in the environmental, social or economic spheres.

The purpose of this study was to carry out a bibliographical and documentary survey of tourism and seasonality and, in an in-depth way, to characterize the object of study, namely the municipality of Guaratuba. In this context, the idea was, through field research, to characterize and prove the existence of seasonality, its causes and effects in the municipality in question. This study was also able to prove the clear existence of second-home tourism in the municipality, a characteristic of many coastal towns. This work was an initial study into the existence of seasonality in the municipality of Guaratuba, which could in future trigger a more qualitative study into the meaning of seasonality from the visitor's point of view and the impact of correlations between different destination contexts and tourists' seasonal experiences, perceptions and constraints.

REFERENCES

ALMEIDA, E. M. de. Coastal tourism: urban occupation and the formation of heat islands. **Tourism in Analysis**. Sao Paulo: USP School of Communication and Arts, volume 11, n. 2, p. 71-89, 2000.

ALMEIDA, A. L. de; KASTENHOLZ E. **Seasonality in tourism and the strategy of diversifying demand**: The case of Northern Portugal. Available at: <http://www.atsie.com/Portals/4/artigos/A%20sazonalidade%20no%20Turismo.pd> Accessed on: 15\06\09.

ANGULO, R.J. As Praias do Paranâ: problemas decorrentes de uma ocupação inadequada. **Revista Paranaense de Desenvolvimento**. Curitiba, n. 99, p. 97-103, Jul./Dec., 2000. Available at:

<http://www.ipardes.gov.br/pdf/revista_PR/99/rodolfo.pdf> Accessed on 10/03/2010.

ASSIS, L.F. de. Second home tourism: the spatial expression of the phenomenon and the possibilities for geographical analysis. **Revista Territòrio**, ano VII, n.11, 12 e 13, p. 107-122, set./out., 2003.

BAHL, M. **Ethnic legacies in the city of Curitiba**: towards the diversification of the local tourist offer. 215 f. Dissertation (Master's Degree in Communication Sciences) Department of Public Relations, Advertising and Tourism, School of Communications and Arts, University of São Paulo - USP. São Paulo, 1994.

M. **Agrupamentos Turisticos Municipais**. Curitiba: Protexto, 2004.

, M. **Viagens e Roteiros Turisticos**. Curitiba: Protexto, 2004.

BARRETTO, M. **Planejamento e Organizaçao do Turismo**. 4. ed. Campinas: Papirus, 1991.

BARROS, N. C. C. de. **Manual** de **geografia do turismo**: meio ambiente, cultura e paisagens. Recife: UFPE University, 1998.

BECKER, B. K. **Survey and evaluation of federal tourism policy and its impact on the coastal region.** Brasilia: Ministry of the Environment, Water Resources and the Legal Amazon, 1995.

BENI, M. **Anàlise Estrutural do Turismo.** Sao Paulo: Senac, 1998.

BONFIM, B. B. R. **Geography in the Training of Tourism Professionals**. 2007. 274f. Thesis (Doctorate in Geography) Faculty of Philosophy, Letters, Human Sciences, University of Sao Paulo, Sao Paulo, 2007.

BOTE GOMÉZ, V. **Planificación económica del turismo:** de una estrategia masiva a una estrategia artesanal. Mexico: Trillas, 1990.

BOULLÓN, R. **Planificación del Espacio Turistico.** 3. ed. México: Trillas, 1997.

CARA, R. B. El turismo y los processos de transformacion territorial. In: RODRIGUES, A. A. B. **Turismo e Geografia - reflexoes teóricas e enfoques regionais.** 3. ed. Sao Paulo: Hucitec, p.86-93, 2001.

CASTELLI, G. **Turismo: Atividade marcante do século XX**. 3. ed. Caxias do Sul: Educs, 1996.

COASTAL COUNCIL. **Integrated Development Master Plan - Guaratuba.** Curitiba: Paranà State Government, 2002.

CORIOLANO, L. N. M. T.; SILVA, S. C. B. de M. e**. Turismo e geografia:** abordagens criticas. Fortaleza: UECE, 2005.

CORIOLANO, L. N. M. Turismo: pràtica social de apropriaçao e de dominaçao de territórios. In publication: **Latin America**: **city, countryside and tourism**. Amalia Inés Geraiges de Lemos, Mónica Arroyo, Maria Laura Silveira. CLACSO, Consejo Latinoamericano de Ciencias Sociales, San Pablo. Diciembre, p. 367-378, 2006. Available at:

<http://bibliotecavirtual.clacso.org.ar/ar/libros/edicion/lemos/21coriol.pdf> Accessed on 10/12/2009.

CORRÊA, R. L. **Regiao e Organizaçao Espacial.** 7. ed. Sao Paulo: Atica, 2000.

CROCIA, N. Regional Analysis and Tourist Destinations: theoretical possibilities and empirical situations in tourism geography. **Turismo - Visao e Açao,** vol. 4 . no. 11, p. 9-28. Available at:

<https://www6.univali.br/seer/index.php/rtva/article/viewFile/1164/920> Accessed on 10/02/2010.

DENCKER, A. de F. M. **Métodos e técnicas** de **pesquisa em turismo.** 3. ed. Sao Paulo: Futura, 1998.

DIAS, R. **Sustainable Tourism and the Environment.** Sao Paulo: Atlas, 2003.

FALCĂO, J. A. G. O turismo internacional e os mecanismos de circulação e transferência de renda. In: Eduardo Yasigi (org.) **Turismo. Space, landscape and culture.** Sao Paulo: Hucitec, 1996.

FRATUCCI, A.C. Tourist places: territories of the tourist phenomenon. **GEOgraphia**, vol. 2, no. 4, p. 121-133, 2000. Available at: <http://www.uff.br/geographia/ojs/index.php/geographia/article/view/45> Accessed on 05/02/2010.

GIL, A. C. **Métodos e Técnicas de Pesquisa Social.** 6. ed. Sao Paulo, Atlas, 2008.

GUARATUBA, Municipal Directorate of Culture, 2010.

GUARATUBA, Municipal Tourism Board, 2009 and 2010.

GUARATUBA, Commercial Association. June 2010.

GUARATUBA, City Hall. 2009. Available at: <http://www.guaratuba.pr.gov.br/site/index.php> Accessed between October 2007 and June 2010.

GUARATUBA, Municipal Department of the Environment. 2009.

ENVIRONMENTAL INSTITUTE OF PARANÁ - IAP. Available at:

<http://www.iap.pr.gov.br/> Accessed May 2010.

BRAZILIAN INSTITUTE OF GEOGRAPHY AND STATISTICS - IBGE. Available at <http://www.ibge.gov.br/home/> Accessed between September 2007 and August 2009.

BRAZILIAN INSTITUTE FOR THE ENVIRONMENT AND RENEWABLE RESOURCES - IBAMA. Available at: <http://ibama.gov.br> Accessed on 10/03/2010.

CHICO MENDES INSTITUTE FOR THE CONSERVATION OF BIODIVERSITY - ICMBio. Available at: <http://icmbio.gov.br> Accessed on 10/03/2010.

PARANÁ INSTITUTE FOR ECONOMIC AND SOCIAL DEVELOPMENT - IPARDES. Available at: <http://www.ipardes.gov.br/index.php> Accessed between August 2008 and May 2010.

PARANAENSE INSTITUTE FOR ECONOMIC AND SOCIAL DEVELOPMENT - IPARDES. **Tourism Production Chain in Paranâ.** Available at: <http://www.ipardes.gov.br/biblioteca/docs/Turismo_Litoral_Parana.pdf> Accessed on May 20, 2009.

IVARS, J. A. Planificación turistica de los espacios regionales en España. Madrid: Sintesis, 2002.

KNAFOU, R. Turismo e Território - por uma abordagem cientifica do turismo. In: RODRIGUES, A.B. **Turismo e Geografia - reflexões teóricas e enfoques regionais**. 3. ed. Sao Paulo: Hucitec, p.62-74, 2001.

KOTLER, P. **Marketing Management:** Analysis, Planning, Implementation and Control. Trad. Ailton Bonfim Brandao. 4. ed. Sao Paulo: Atlaş, 1994.

KRIPPENDORF, J. **Sociologia do Turismo:** para uma nova compreensão do lazer e das viagens. Rio de Janeiro: Civilizaçao Brasileira, 1989.

. **Sociology of Tourism:** towards a new understanding of leisure and travel. Sao Paulo: Aleph, 2000.

LAGE, H. G.; MILONE, P. C. **Economia do turismo.** Campinas: Papirus, 1998.

. Global socio-economic impacts of tourism. In:

LAGE, H. G.; MILONE, P. C. **Turismo - teoria e pràtica.** Sao Paulo: Atlas, p. 117131, 2000.

LUCAS, O. W. R. **The Design of Forest Landscape**. New York: Oxford University Press. 1991.

LUCHIARI, M. T. D. P. Tourism and the Environment in the Mitification of Places. **Tourism in Analysis**. Sao Paulo: USP School of Communication and Arts, volume 11, n. 1, p. 35-43, 2000.

MAFRA, J. S. **History of the Municipality of Guaratuba**. Guaratuba, 1952.

MARCONI, M. de A.; LAKATOS, E. M. **Técnicas de Pesquisa.** 7. ed. Sao Paulo: Atlas, 2008.

MENDONÇA, F; KOZEL, S. (eds) **Elements of Epistemology of Contemporary Geography.** Curitiba: UPFR, 2004.

MOLINA, F. S. **Tourism and the Production of Space:** The Case of Jericoacoara, CE. 2007.150f. Dissertation (Master's Degree in Human Geography) Faculty of Philosophy, Letters, Human Sciences, University of Sao Paulo, Sao Paulo, 2007.

MONTEJANO, J. **Estrutura del Mercado Turistico**. Madrid: Sintesis, 1996.

MORAES, A. C. R. **Território e história no Brasil**. Sao Paulo: Hucitec, 2002.

MOTA, K. C. N. **Marketing Turistico:** promovendo uma atividade sazonal. Sao Paulo: Atlas, 2001.

MOURA, R.; WERNECK, D.Z. Paranà Coastal Continuous Occupation: a reading of space. **Revista Paranaense de Desenvolvimento**. Curitiba, n. 99, p. 61-82, Jul./Dec., 2000. Available at:<http://www.ipardes.gov.br/pdf/revista_PR/99/rodolfo.pdf> Accessed on March 10, 2010.

NICOLAS, D. H. Elementos para un anàlisisis sociogeogràfico del Turismo. In:

Rodrigues A.B. (org.) **Turismo e Geografia - Reflexoes Teóricas e Enfoques Regionais.** Sao Paulo: Hucitec, 1996.

UNWTO. **What every tourism manager should know.** Madrid (Spain): UNWTO, 1995.

UNWTO. **Introduction to tourism**. Trad. Dolores Martin Rodriguez Corner. Sao Paulo: Roca. 2001.

PARANA, Secretary of State for Tourism. **Tourist Region:** Litoral do Paranà in data, 2008. Available at: <http://www.setu.pr.gov.br> Accessed on 05/07/2009.

PARANA, State Secretariat for Tourism. Available at:

<http://www.setu.pr.gov.br/> Accessed between August 2007 and August 2009.

PARANAGUA, City Hall. 2010. Available at:

<http://www.paranagua.pr.gov.br/> Accessed between February and May 2010.

PEARCE, D. **Tourism Today** - A Geographical Analysis. New York: Longman, 1991.

PETROCCHI, M. **Turismo**: Planejamento e Gestao. 2. ed. Sao Paulo: Futura, 1998.

PIERRI, N; *et al.* Occupation and land use on the coast of Paraná: constraints, conflicts and trends. In: **Desenvolvimento e Meio Ambiente.** Curitiba, Editora UFPR, n. 13, p. 137-167, jan/jun, 2006.

PONTAL DO PARANA, City Hall. **Urban Development Master Plan.** 2. ed. Primeiro Caderno, Pontal do Paranà, 2001.

REJOWSKI, M. **Turismo e pesquisa cientifica**: pensamento internacional x situaçao brasileira. Campinas: Papirus, 1996 (Tourism Collection).

RODRIGUES, A. B. **Turismo e Espaço.** Towards transdisciplinary knowledge. Sao Paulo: HUCITEC, 1997.

RUARO, D. A. **Manual de Apresentação de Produçao Acadêmica.** 2. ed. Pato Branco, Faculdade Mater Dei, 2004.

RUSCHMANN, D. **Marketing Turistico:** Um enfoque promotional. 2. ed. Campinas: Papirus, 1995.

SAMPAIO, R. Occupation of Paraná's seashores by beach use. In: **Desenvolvimento e Meio Ambiente.** Curitiba, Editora UFPR, n. 13, p. 169-186, jan/jun, 2006.

SANTOS, M. **Espaço e sociedade.** Petrópolis: Vozes, 1979.

. **Metamorfoses do espaço habitado.** 2. ed. Sao Paulo: Hucitec, 1991.

. **Technique, space, time, globalization and the technical-scientific informational environment**. Sao Paulo: Hucitec, 1994.

. **Space & Method**. 4. ed. Sao Paulo: Nobel, 1997.

SCHEUER, L. **Tourism Development Plan for the Municipality of Guaratuba**.

AMLIPA, 2010.

SECRETARY OF STATE FOR THE ENVIRONMENT - SEMA. **Guaratuba Environmental Protection Area**. Curitiba, 2001.

SEVERINO, A. J. **Metodologia do trabalho cientifico.** 20. ed. Sao Paulo: Cortez, 1996.

SILVEIRA, M. A. T. da. **Tourism, land-use planning and development policies.** A focus on the state of Paranâ in the regional context. 277 f. Thesis (Doctorate in Geography) Faculty of Philosophy, Letters, Human Sciences, University of Sao Paulo - USP, Sao Paulo,

2002.

. **The Influence of Seasonality on the Development of Tourist Destinations:** A Contribution to the Theme. 2010, in press.

SONEIRO, J. C. **Aproximación a la geografia del turismo**. Madrid: Sintesis, 1991.

SOUZA, M. J. O. Território: sobre espaço e poder, autonomia e desenvolvimento. In: Ina Elias de Castro et all (org.). **Geografia: conceitos e temas**. Rio de Janeiro: Bertrand Brasil, 1995.

TABARES, F. C. **Produto Turistico:** aplicación de la estadistica y del muestro para sua deseno. 2.ed. Mexico: Trillas, 1991.

TRIGO, L. G. **Turismo e Qualidade: Tendências Contemporâneas**. 6 ed. Campinas: Papirus, 1993.

TULIK, O. **Secondary residences**: presence, size and expressiveness of the phenomenon in the state of São Paulo. 154 f. Thesis (Livre-Docência) Faculdade de Filosofia, Letras, Ciências Humanas, Universidade de Sao Paulo - USP, Sao Paulo, 1995.

VAZ, G.N. **Marketing Turistico:** receptivo e emissivo. Sao Paulo: Pioneira, 1999.

VERA, J. F. **Analisis territorial del turismo**. Barcelona: Ariel, 1997.

WAHAB, S.E.A. **Introduction to Tourism Management:** some structural and operational aspects of international tourism - theory and practice. 3. ed. Sao Paulo: Pioneira, 1991.

WORD TOURISM ORGANIZATION. WTO/OMT. <http://www.unwto.org/index.php> Accessed between September 2007 and March 2010.

MIX
Papier aus verantwortungsvollen Quellen
Paper from responsible sources
FSC® C105338

Printed by Books on Demand GmbH, Norderstedt / Germany